Concepts of Semiconductor Photocatalysis

Edited by Mohammed Rahman, Anish Khan, Abdullah Asiri and Inamuddin Inamuddin

Published in London, United Kingdom

IntechOpen

Supporting open minds since 2005

Concepts of Semiconductor Photocatalysis
http://dx.doi.org/10.5772/intechopen.77477
Edited by Mohammed Rahman, Anish Khan, Abdullah Asiri and Inamuddin Inamuddin

Contributors
Naorem Khelchand Singh, Rajshree Rajkumari, Keyla M. Fuentes, Sara Aldabe Bilmes, Margarita Sanchez-Dominguez, Sahar Zinatloo_Ajabshir, Zahra Sayyar, Jamshaid Rashid, Fatima Imtiaz, Ming Xu, Yuichi Negishi, Wataru Kurashige, Sakiat Hossain, Marika Aoki, Seiichiro Watanabe, Manuel Alberto Flores-Hidalgo, Diana Barraza-Jimenez, Sandra Iliana Torres-Herrera, Elva Marcela Coria-Quiñones, Raúl Armando Olvera-Corral, David José Romero-Soto, Adolfo Ruiz Soto, Mohammed Muzibur Rahman

First published in London, United Kingdom, 2019 by IntechOpen
IntechOpen is the global imprint of INTECHOPEN LIMITED, registered in England and Wales, registration number: 11086078, 7th floor, 10 Lower Thames Street, London, EC3R 6AF, United Kingdom
Printed in Croatia

British Library Cataloguing-in-Publication Data
A catalogue record for this book is available from the British Library

Additional hard and PDF copies can be obtained from orders@intechopen.com

Concepts of Semiconductor Photocatalysis
Edited by Mohammed Rahman, Anish Khan, Abdullah Asiri and Inamuddin Inamuddin
p. cm.
Print ISBN 978-1-78985-087-1
Online ISBN 978-1-78985-088-8
eBook (PDF) ISBN 978-1-83968-245-2

Meet the editors

Mohammed Muzibur Rahman: He received his B.Sc. and M.Sc. from Shahjalal University of Science & Technology, Sylhet, Bangladesh in 1999 and 2001 respectively. He received his Ph.D. from the Chonbuk National University, South Korea, in 2007. After his Ph.D., he worked in a postdoctoral fellowship and as an Assistant Professor in pioneer research centers and universities located in South Korea, Japan, and Saudi Arabia (2007 to 2011). Presently, he is working as an Associate Professor in the Center of Excellence for Advanced Materials Research (CEAMR) and the Chemistry Department at King Abdulaziz University, Saudi Arabia since 2011. He has published more than 250 papers for international and domestic conferences, published several book chapters, and edited ten books. His research work has been largely in the area of photocatalysis, semiconductor, nanoparticles, carbon nanotubes, nanotechnology, electrocatalysis, sensors, ionic liquid, surface chemistry, gold nanoparticles, electrochemistry, instrumental science, nanomaterials, self-assembled monolayers, photochemistry, m-chips, and devices etc.

Abdullah Mohamed Asiri: He received his PhD from the University of Wales, College of Cardiff, UK in 1995. He has been the Head of the Chemistry Department at King Abdulaziz University since October 2009 and he is the founder and the Director of the Center of Excellence for Advanced Materials Research (CEAMR). He is a Professor of Organic Photochemistry. His research interests cover color chemistry, synthesis of novel photochromic and thermochromic systems, synthesis of novel coloring matters and dyeing of textiles, materials chemistry, nanochemistry, nanotechnology, polymers, and plastics. He is the Editor-in-Chief of King Abdul Aziz University Journal of Science. He is also a member of the Editorial Board of Pigments and Resin Technology (UK), Organic Chemistry in Sight (New Zealand), and Recent Patents on Materials Science (USA). He is the Vice-President of the Saudi Chemical Society (Western Province Branch).

Dr. Inamuddin: He is currently working as an Assistant Professor in the Chemistry Department, Faculty of Science, King Abdulaziz University, Jeddah, Saudi Arabia. He has extensive research experience in the multidisciplinary fields of analytical chemistry, materials chemistry, and electrochemistry and, more specifically, renewable energy and the environment. He has published 127 research articles in international journals of repute and eighteen book chapters in knowledge-based book editions published by renowned international publishers. He has published thirty-nine edited books with Springer, United Kingdom, Elsevier, Nova Science Publishers, Inc. U.S.A., CRC Press Taylor & Francis Asia Pacific, Trans Tech Publications Ltd., Switzerland,

and Materials Science Forum, U.S.A. He is a member of various editorial boards of journals and serves as an associate editor for journals such as Environmental Chemistry Letter, Applied Water Science, Euro-Mediterranean Journal for Environmental Integration, Springer-Nature, editorial board member for Scientific Reports-Nature and editor for Eurasian Journal of Analytical Chemistry.

Dr. Anish Khan : He is currently working as an Assistant Professor, Chemistry Department, Centre of Excellence for Advanced Materials Research (CEAMR), Faculty of Science, King Abdulaziz University, Jeddah, Saudi Arabia. He received his Ph.D. from Aligarh Muslim University, India in 2010. He has research experience in the field of synthetic polymers and organic-inorganic electrically conducting nano-composites. He completed his postdoctoral studies from the School of Chemical Sciences, University Sains Malaysia (USM) in electroanalytical chemistry in 2010-2011. He has research and teaching experience and has published more than 127 research papers in international journals. He has attended more than 20 international conferences and workshops and has published 5 books with 14 books currently in progress and 14 book chapters. Preparation and characterization of hybrid nano composite materials and their applications, Polymeric inorganic cation –exchange materials, Electrically conducting polymeric, materials, Composite material use as Sensors, Green chemistry by remediation of pollution, Heavy metal ion selective membrane electrode, Biosensor on neurotransmitter.

Contents

Preface

This book "**Concepts of Semiconductor Photocatalysis**" contains recent research on the preparation, characterization, and potential applications of the photocatalyst. The book includes information on solar exposure of the photocatalyst to ultraviolet light, which affects redox reactions, degradation, molecular transformations, as well as colored material mineralization. The development of new and efficient technology is pointing researchers toward the safe, facile, non-toxic, and eco-friendly route of synthesis-to-applications, which can be used for manufacture at a large scale. This is a simple, cost-effective, stable for a long time, and reproducible aqueous room temperature synthesis method to obtain an efficient semiconductor photocatalyst. This potentially unique work offers various approaches on R&D with semiconductor photocatalyst materials in an aqueous or non-aqueous phase through fully modified or unmodified states. We know, recently the nanotechnology offer the regulating of semiconductor substances at the nano-scale and nano-dimensional substances in few nano-meters, which exclusive phenomenon facilitates us to control the novel catalytic applications with photocatalysts. Photocatalysis reactions performed in the presence of a low-dimensional semiconductor and light are rapidly becoming one of the most dynamic areas of catalysis research, with potential applications in areas such as environmental, solar, renewable energy, medicine, and nanomaterials. This was investigated in this book using the techniques of photocatalyst preparation, various types of characterization, and possible industrial applications related to solar photocatalyst research.

Mohammed Muzibur Rahman
Center of Excellence for Advanced Materials Research (CEAMR)
and Chemistry Department,
Faculty of Science,
King Abdulaziz University,
Jeddah, Saudi Arabia

Chapter 1

Introductory Chapter: Fundamentals of Semiconductor Photocatalysis

Mohammed Muzibur Rahman

1. Introduction

In semiconductor photo-catalysis, generally it is used to design, characterize, and potential photocatalytical applications with doped or undoped nanostructural materials especially considering their particle sizes as well as shapes. Nanostructured materials, for example, porous metal oxides, metal nanoparticles, porous carbons, and their composites are widely studied for their potential applications in energy conversion/storage devices, gas storage, photocatalysis as well as in electrocatalysis. Photocatalysis is the phenomena, which occurs in catalysis by influence of photons. An efficient photocatalyst is a conductive nanomaterial, which directly absorbs incident light to bring up to higher energy states, which provides such energy to a reacting substance to make a chemical reaction occur. The mechanism of *photocatalytic* water splitting over *semiconductor photocatalysts* is also discussed in this chapter. Thermodynamically, the reaction of complete water splitting under normal conditions is a highly endothermic process because the change in the standard Gibbs energy is very large.

2. Literature review

For photocatalytic applications, the location of the TiO_2 nanoparticle plays an important role in the photoactivity, since various nanoparticles should be accessible for both photons and reactive molecules. The advantages of using these materials as support for TiO_2 in relation to photocatalytic processes are addressed by comparing the photocatalytic activity of TiO_2 nanoparticles deposited by different routes with TiO_2 nanoparticles on mesoporous materials [1–3]. Three-dimensional nanostructures have interconnected networks of bulk materials and larger accessible surface areas with respect to their lower dimensional counterparts. The surface functionalities, controllable host-guest interactions, and size-selective sensing properties are enabled due to the spatial arrangement of particles within three-dimensional framework [4–15].

Here, catalysts based on Ti-containing mesoporous silica have been developed from different synthetic pathways for selective oxidation of bulky organic compounds and photoreduction of greenhouse gases [16–18]. To support particles of TiO_2 on mesoporous silica, maintaining or improving its photocatalytic activity is not an easy task as, in addition to be strongly attached to the substrate for avoiding leakage, nanoparticles' size, location, and agglomeration should be controlled to have accessibility for adsorbates and incident photons [19–22]. Nanostructure of rare earth material is appropriate for these applications that can be synthesized with different methods, for example co-precipitation, hydrothermal, sol–gel, combustion, stearic acid route, pechini, cathode plasma, electrolysis, co-ions complexation, molten salt, and other approaches [13]. Rare-earth-based materials were used in different fields. In this section, we study a brief review of rare-earth-doped materials, rare-earth-based oxide–oxide composites, metal-modified semiconductors, and mixed-oxide materials in the fields of photocatalytic application as solar energy generation. When a photon is absorbed by rare-earth-based material using the energy gap of the materials, electron–hole pairs are generated in the photocatalysis mechanism [23–27]. Semiconductor photocatalysis can be of two types, homogenous photocatalysis and heterogeneous photocatalysis, depending on the phase differences of reactants and catalysts, which triggered the concept of advanced oxidation processes at the end of twentieth century. TiO_2 is the semiconductor, which can act as both the hetero- and homogenous photocatalyst photoreactions. There are numerous semiconductors, which can be coupled with other semiconductors and can be incorporated into heterostructures or composites. So, another example of inorganic heterostructure is the template-free simple synthesis of CdS-ZnO nanocomposites. This self-assembled flower-like structure resulted from coupling of two semiconductors, which increased the charge separation and demonstrated greater photocatalytic activity [28, 29].

Negheshi et al. attempted to improve the activity of mesoporous tungsten trioxide, and titanium dioxide (m-WO_3 and m-TiO_2) photocatalysts, which convert methane into methanol, by loading the ultrafine metal clusters as co-catalyst on the photocatalysts. They have succeeded in loading ultrafine metal-cluster co-catalysts onto m-WO_3 and m-TiO_2, and thereby improving their photocatalytic activity in presence of mesoporous nanomaterials. The photocatalytic activity measurements clearly demonstrated that the loading of such co-catalysts is effective in improving the activity for both types of photocatalysts. It is important to appropriately select the element of the co-catalyst according to the photocatalyst to improve the photocatalytic activity. In this approach, the particle diameter of the co-catalyst was reduced to approximately 1.0 nm to increase the reactive surface area [29–32]. Textile dyes and other industrial dyestuffs constitute one of the largest groups of organic compounds that signify enhancing environmental pollution.

Contaminated water means undesirable substances; it adversely affects the quality and makes it inappropriate for use. Various sources such as residential, commercial, industrial practices, etc. contaminate the water and its sources as well. Wastewater parameters vary widely and depend on the source it is produced form. They are usually pathogenic, nonpathogenic microorganisms, organic, or inorganic [33, 34]. The toxic and carcinogenic nature of these dyes and their manufacturing precursors represent a danger to human beings and perturbation in aquatic life. The enlarged public concern about these environmental pollutants and the creation of international environmental standards have prompted the need for the development of novel treatment methods for converting organic pollutants (such as dye effluents) to harmless compounds. Reductive cleavage via anaerobic biological treatment generates potentially carcinogenic intermediates or end products, especially aromatic amines. Physical processes are like reverse osmosis, flocculation, absorption, air-stripping, combustion, and aerobic biological oxidation, which are nondestructive, and these processes only shift the pollutants from one medium to others, thus creating secondary pollution [35–38]. For the past two decades, a lot of research is going on for advanced oxidation processes, pointing out its potential vital role in the wastewater purification, in which the high oxidizing potential species like hydroxyl radicals are utilized as an alternative way of treating undesirable organic pollutants. Advanced oxidation processes shows great effort in solving the bio-recalcitrant water-pollutant issues at near-ambient temperature and pressure [39–41]. Heterogeneous photocatalysis is a method wherein a blend of photochemistry and catalysis is operable and suggests that light and catalyst are necessary to bring out a chemical reaction. Lately, there has been a developing enthusiasm in usage of semiconductors as photosensitizers for the complete oxidative mineralization of pollutants by oxygen [42, 43]. Heterogeneously scattered semiconductor surfaces give both a fixed situation to impact the chemical reactivity of a wide scope of adsorbates and a means to initiate light-induced redox reactivity in these feebly related molecules.

When a semiconductor is intact with an aqueous solution, thermodynamic equilibration takes place at the interface. This may result in the development of a space-charge layer inside a slight surface area of the semiconductor, in which the electronic vitality bands are commonly twisted downward or upward, respectively, in the instances of p- and n-type semiconductors. The space-charge layer thickness is typically in the range of 1–10^3 nm, contingent upon the dielectric constant and carrier density of the semiconductor. If this semiconductor perceives photons with energies that are more noteworthy than that of the respective material's band gap energy, electron–hole pairs are produced and parted in the space-charge layer. The development of new and noble nanotechnology is urgently required for introducing the eco-friendly, safe, facile, non-toxic routes in synthesis-to-applications, which can be used by industrial sectors in broad scale.

Recently, the nanotechnology offers the control growth of semiconductor substances at the nanoscale, which exclusively facilitates the novel photocatalytic applications with transition doped semiconductor catalysts. Photocatalysis, reactions supported out in the presence of a low-dimensional semiconductor and light, is rapidly becoming one of the most dynamic areas of catalysis research, with potential applications in areas such as environmental, solar, renewable energy, medicine, and sensors [44–47]. Here, in this book, it presents an overview of current photocatalyst fundamental, substantial applications, and enactment of the research in worldwide. This investigated the techniques of photocatalyst preparation, various types of characterization, and possible industrial applications related to solar photocatalyst researches.

Author details

Mohammed Muzibur Rahman
Center of Excellence for Advanced Materials Research (CEAMR) and Chemistry Department, Faculty of Science, King Abdulaziz University, Jeddah, Saudi Arabia

*Address all correspondence to: mmrahmanh@gmail.com

References

[1] Gustafsson H, Isaksson S, Altskär A, Holmberg K. Mesoporous silica nanoparticles with controllable morphology prepared from oil-in-water emulsions. Journal of Colloid and Interface Science. 2016;**467**:253-260

[2] Subhan MA, Jhuma SS, Saha PC, Alam MM, Asiri AM, Al-Mamun M, et al. Efficient selective 4-aminophenol sensing and antibacterial activity of ternary $Ag_2O_3{\cdot}SnO_2{\cdot}Cr_2O_3$ nanoparticles. New Journal of Chemistry. 2019;**43**:10352-10365

[3] Wahid A, Asiri AM, Rahman MM. One-step facile synthesis of Nd_2O_3/ZnO nanorods for an efficient selective 2,4-dinitrophenol sensor probe. Applied Surface Science. 2019;**487**:1253-1261

[4] Rahman MM, Sheikh TA, Asiri AM, Awual MR. Development of 3-methoxyanaline sensor probe based on thin $Ag_2O@La_2O_3$ nanosheets for environmental safety. New Journal of Chemistry. 2019;**43**:4620-4632

[5] Rahman MM, Wahid A, Alam MM, Asiri AM. Efficient 4-Nitrophenol sensor development based on facile $Ag@Nd_2O_3$ nanoparticles. Materials Today Communications. 2018;**16**:307-313

[6] Rahman MM, Alam MM, Asiri AM. Sensitive 1,2-dichlorobenzene chemi-sensor development based on solvothermally prepared FeO/CdO nanocubes for environmental safety. Journal of Industrial and Engineering Chemistry. 2018;**62**:392-400

[7] Rahman MM. A label-free kanamycin sensor development based on CuO-NiO hollow-spheres: Food samples analyses. Sensors & Actuators: B Chemical. 2018;**264**:84-91

[8] Rahman MM, Alam MM, Asiri AM. 2-Nitrophenol sensor-based wet-chemically prepared binary doped Co_3O_4/Al_2O_3 nanosheets by an electrochemical approach. RSC Advances. 2018;**8**:960-970

[9] Rahman MM, Ahmed J, Asiri AM. Thiourea sensor development based on hydrothermally prepared CMO nanoparticles for environmental safety. Biosensors and Bioelectronics. 2018;**99**:586-592

[10] Ahmed J, Rahman MM, Siddiquey IA, Asiri AM, Hasnat MA. Efficient hydroquinone sensor based on zinc, strontium and nickel based ternary metal oxide (TMO) composites by differential pulse voltammetry. Sensors and Actuators B: Chemical. 2018;**256**:383-392

[11] Hu Y, Martra G, Zhang J, Higashimoto S, Coluccia S, Anpo M. Characterization of the local structures of Ti-MCM-41 and their photocatalytic reactivity for the decomposition of NO into N_2 and O_2. The Journal of Physical Chemistry. B. 2006;**110**:1680-1685

[12] Li J, Zhou C, Xie H, Ge Z, Yuan L, Li X. Titanium-containing mesoporous materials: Synthesis and application in selective catalytic oxidation. Journal of Natural Gas Chemistry. 2006;**15**:164-177

[13] Hossain MA, Elias M, Sarker DR, Diba ZR, Mithun JM, Azad MAK, et al. Synthesis of Fe- or Ag-doped TiO_2–MWCNT nanocomposite thin films and their visible-light-induced catalysis of dye degradation and antibacterial activity. Research on Chemical Intermediates. 2018;**44**:2667-2683

[14] Rahman MM, Asiri AM, Youssef TE, Marwani HM. Photocatalytic degradation of Remazol Brilliant Orange 3R using wet-chemically prepared CdO-ZnO nanofibers for environmental remediation. Material Express. 2016;**6**:137-148

[15] Zinatloo-Ajabshir S, Salavati-Niasari M, Sobhani A, Zinatloo-Ajabshir Z. Rare earth zirconate nanostructures: Recent development on preparation and photocatalytic applications. Journal of Alloys and Compounds. 2018;**767**:1164-1185

[16] Liu Y, Deng Y, Sun Z, Wei J, Zheng G, Asiri AM, et al. Hierarchical Cu_2S microsponges constructed from nanosheets for efficient photocatalysis. Small. 2013;**9**:2702-2708

[17] Jamal A, Rahman MM, Khan SB, Faisal M, Asiri AM, Khan AAP, et al. Hydrothermally preparation and characterization of un-doped manganese oxide nanostructures: Efficient photocatalysis and chemical sensing applications. Micro and Nanosystems. 2013;**5**:22-28

[18] Ahmed J, Rahman MM, Siddiquey IA, Asiri AM, Hasnat MA. Efficient bisphenol-A detection based on the ternary metal oxide (TMO) composites by electrochemical approaches. Electrochimica Acta. 2017;**246**:597-605

[19] Rahman MM, Ahmed J. Cd-doped Sb_2O_4 nanostructures modified glassy carbon electrode for efficient detection of melamine by electrochemical approach. Biosensors and Bioelectronics. 2018;**102**:631-636

[20] Subhan MA, Saha PC, Rahman MM, Ahmed J, Asiri AM, Al-Mamun M. Fabrication of a 2,4-dinitrophenol sensor based on Fe_3O_4@Ag@Ni nanomaterials and studies of their antibacterial properties. New Journal of Chemistry. 2018;**42**:872-881

[21] Alam MM, Asiri AM, Uddin MT, Inamuddin MAI, Awual MR, Rahman MM. One-step wet-chemical synthesis of ternary ZnO/CuO/Co_3O_4 nanoparticles for sensitive and selective melamine sensor development. New Journal of Chemistry. 2019;**43**:4849-4858

[22] Rahman MM, Alam MM, Asiri AM. Efficient phenolic sensor development based on facile Ag_2O/Sb_2O_3 nanoparticles for environmental safety. Nanoscale Advances. 2019;**1**:696-705

[23] Hussein MA, Alam MM, Alenazi NA, Alamry KA, Asiri AM, Rahman MM. Nanocomposite based functionalized polyethersulfone and conjugated ternary ZnYCdO nanomaterials for the fabrication of selective Cd^{2+} sensor probe. Journal of Polymer Research. 2018;**25**:262

[24] Sheikh TA, Rahman MM, Asiri AM, Marwani HM, Awual MR. 4-Hexylresorcinol sensor development based on wet-chemically prepared Co_3O_4@Er_2O_3 nanorods: A practical approach. Journal of Industrial & Engineering Chemistry. 2018;**66**:446-455

[25] Rahman MM, Alam MM, Asiri AM. Selective hydrazine sensor fabrication with facile low-dimensional Fe_2O_3/CeO_2 nanocubes. New Journal of Chemistry. 2018;**42**:10263-10270

[26] Awual MR, Hasan MM, Islam A, Rahman MM, Asiri AM, Khaleque MA, et al. Offering an innovative composited material for effective lead(II) monitoring and removal from polluted water. Journal of Cleaner Production. 2019;**231**:214-223

[27] Awual MR, Hasan MM, Asiri AM, Rahman MM. Cleaning the arsenic(V) contaminated water for safe-guarding the public health using novel composite material. Composites Part B: Engineering. 2019;**171**:294-301

[28] Awual MR, Hasan MM, Islam A, Rahman MM, Asiri AM, Khaleque MA, et al. Introducing an amine functionalized novel conjugate

material for toxic nitrite detection and adsorption from wastewater. Journal of Cleaner Production. 2019;**228**:778-785

[29] Awual MR, Islam A, Hasan MM, Rahman MM, Asiri AM, Khaleque MA, et al. Introducing an alternate conjugated material for enhanced lead(II) capturing from wastewater. Journal of Cleaner Production. 2019;**224**:920-929

[30] Awual MR, Hasan MM, Rahman MM, Asiri AM. Novel composite material for selective copper(II) detection and removal from aqueous media. Journal of Molecular Liquids. 2019;**283**:772-780

[31] Awual MR, Hasan MM, Asiri AM, Rahman MM. Novel optical composite material for efficient vanadium(III) capturing from wastewater. Journal of Molecular Liquids. 2019;**283**:704-712

[32] Rao GV, Sree GS, Mohan BS, Anjaneyulu RB. A simple and green synthesis of Zinc oxide nanoparticles using *Gloriosa superba* leaves extract. International Journal of Green and Herbal Chemistry. 2018;7(2):373-378

[33] Robertson P. Semiconductor photocatalysis: An environmentally acceptable alternative production technique and effluent treatment process. Journal of Cleaner Production. 1996;**1996**(4):203

[34] Turchi CS, Ollis DF. Photocatalytic degradation of organic water contaminants: Mechanisms involving hydroxyl radical attack. Journal of Catalysis. 1990;**122**:178

[35] Awual MR, Khraisheh M, Alharthi NH, Luqman M, Islam A, Karim MR, et al. Efficient detection and adsorption of cadmium(II) ions using innovative nano-composite materials. Chemical Engineering Journal. 2018;**343**:118-127

[36] Awual MR, Alharthi NH, Hasan MM, Karim MR, Islam A, Znad H, et al. Inorganic-organic based novel nano-conjugate material for effective cobalt(II) ions capturing from wastewater. Chemical Engineering Journal. 2017;**324**:130-139

[37] Awual MR, Hasan MM, Eldesoky GE, Khaleque MA, Rahman MM, Naushad M. Facile mercury detection and removal from aqueous media involving ligand impregnated conjugate nanomaterials. Chemical Engineering Journal. 2016;**290**:243-251

[38] Katowah DF, Rahman MM, Hussein MA, Sobahi TR, Gabal MA, Alam MM, et al. Ternary Nanocomposite based poly(pyrrole-co-O-toluidine), cobalt ferrite and decorated chitosan as a selective Co^{2+} cationic sensor. Composites Part B: Engineering. 2019;**175**:107175

[39] Rahman MM, Alam MM, Alamry KA. Sensitive and selective m-Tolylhydrazine sensor development based on CdO nanoparticles decorated multi-walled carbon nanotubes. Journal of Industrial and Engineering Chemistry. 2019;**77**:309-316

[40] Rahman MM, Hussain MM, Arshad MN, Awual MR, Asiri AM. Arsenic sensor development based on modified with (E)-N′-(2-nitrobenzylidine)-benzenesulfonohydrazide: A real sample analysis. New Journal of Chemistry. 2019;**43**:9066-9075

[41] Abdu-Zied BM, Alam MM, Asiri AM, Schwieger W, Rahman MM. Fabrication of 1,2-dichlorobenzene sensor based on mesoporous MCM-41 material. Colloids and Surfaces A: Physicochemical and Engineering Aspects. 2019;**562**:161-169

[42] Karim MR, Alam MM, Aijaz MO, Asiri AM, Dar MA, Rahman MM. Fabrication of 1,4-dioxane sensor

based on microwave assisted PAni-SiO_2 nanocomposites. Talanta. 2019;**193**:64-69

[43] Alam MK, Rahman MM, Rahman MM, Kim D, Asiri AM, Khan FA. In-situ synthesis of gold nanocrystals anchored graphene oxide and its application in biosensor and chemical sensor. Journal of Electroanalytical Chemistry. 2019;**835**:329-337

[44] Rahman MM, Alam MM, Asiri AM. Carbon black co-adsorbed ZnO nanocomposites for selective benzaldehyde sensor development by electrochemical approach for environmental safety. Journal of Industrial and Engineering Chemistry. 2018;**65**:300-308

[45] El-Shishtawy RM, Al-Ghamdi HA, Alam MM, Al-Amshany ZM, Asiri AM, Rahman MM. Development of Cd^{2+} sensor based on BZNA/nafion/glassy carbon electrode by electrochemical approach. Chemical Engineering Journal. 2018;**352**:225-231

[46] Hussain MM, Asiri AM, Arshad MN, Rahman MM. Fabrication of a Ga^{3+} sensor probe based on methoxy benzylidene benzene sulfo no hydrazine (MBBSH) by an electrochemical approach. New Journal of Chemistry. 2018;**42**:1169-1180

[47] Rahman MM, Alam MM, Asiri AM, Islam MA. 3,4-Diaminotoluene sensor development based on hydrothermally prepared $MnCo_xO_y$ nanoparticles. Talanta. 2018;**176**:17-25

Chapter 2

Semiconductor Nanocomposites for Visible Light Photocatalysis of Water Pollutants

Fatima Imtiaz, Jamshaid Rashid and Ming Xu

Abstract

Semiconductor photocatalysis gained reputation in the early 1970s when Fujishima and Honda revealed the potential of TiO_2 to split water in to hydrogen and oxygen in a photoelectrochemical cell. Their work provided the base for the development of semiconductor photocatalysis for the environmental remediation and energy applications. Photoactivity of some semiconductors was found to be low due to larger band gap energy and higher electron-hole pair recombination rate. To avoid these problems, the development of visible light responsive photocatalytic materials by different approaches, such as metal and/or non-metal doping, co-doping, coupling of semiconductors, composites and heterojunctions materials synthesis has been widely investigated and explored in systematic manner. This chapter emphasizes on the different type of tailored photocatalyst materials having the enhanced visible light absorption properties, lower band gap energy and recombination rate of electron-hole pairs and production of reactive radical species. Visible light active semiconductors for the environmental remediation purposes, particularly for water treatment and disinfection are also discussed in detail. Studies on the photocatalytic degradation of emerging organic compounds like cyanotoxins, VOCs, phenols, pharmaceuticals, etc., by employing variety of modified semiconductors, are summarized, and a mechanistic aspects of the photocatalysis has been discussed.

Keywords: visible light photocatalysis, semiconductor composites, organic pollutants, mechanism, wastewater treatment

1. Introduction to photocatalysis

Catalysis refers to a phenomenon in which a substance (the catalyst) speed up a kinetically slow reaction and the catalyst is fully restored at the end of each catalytic cycle and the "Photocatalysis" is defined as a specific process for the acceleration of the "photoreaction" in the presence of a catalyst [1]. Photocatalysis can also be regarded as the catalysis of photochemical reaction on a solid substrate, mostly a semiconductor [2]. The term "Photocatalysis" is still an arguable subject due to controversies, according to some researchers the "light" acts as a catalyst, while it always acts as a "reactant" where it is spent in the chemical process [3].

The word "photoreaction" is sometimes explained as a "Photoinduced" or "Photoactivated," process, whereas in the field of photocatalysis, "catalytic activity" is the ability of a catalyst to show performance under light depending upon the reaction sites/active sites at the catalyst. The performance of a catalyst can be determined by its "turnover frequency" which is "number of turnovers per unit time of reaction, it is used to show how many times one active site produces a reaction product(s) within unit time." In case of photocatalysis, the reaction rate depends on the frequency of irradiated light which acts as the initiator of photoreaction. The term photocatalysis indicates the relation of light and some substance; (say a catalyst) so in the absence of light, the process of photocatalytic activities on active sites is not possible [1, 4].

1.1 Brief history of semiconductor photocatalysis

In 1839, the production of voltage and an electric current were reported when silver chloride electrode, immersed in electrolyte solution (connected to counter electrode) exhibited illumination under sunlight, which lately was known as "Becquerel effect" led to the beginning of existing era of photo electrochemistry [3]. In 1921, the Renz published first article on the degradation of carbon compounds by using titanium dioxide. In 1924 the photocatalytic deposition of silver on zinc oxide was observed for the production of metallic silver [2]. In 1929, the "chalking" (fading of paints) effect of titanium white (TiO_2), under the strong sunlight was observed, which in 1938 became the basis for self-cleaning property and modification in the concept of photo-induced reduction [5].

During 1950s, the insights of photocatalysis were shifted to ZnO. In 1953, two studies for the production of H_2O_2 on ZnO under the UV irradiation started a series of follow-up studies in the upcoming years which explained that an organic compound is oxidized when atmospheric oxygen is reduced. Till 1955, the photocatalytic behaviour of ZnO, Sb_2O_3 and TiO_2 including photoconductivity and fluorescence were well known. In 1958, the adsorption of reduced O_2 on the TiO_2 surface, as a result of photoexcitation, having the ability to degrade a dye was studied. During 1960s and 1970s, the processes and concepts of photo-oxidation of CN^- and photodeposition of Pt, Cu, Pd, and other metals on TiO_2, WO_3, Al_2O_3, and SnO_2 to be used as co-catalysts, had been evolved. In 1972 the field of photo-electrochemical got more attention when Fujishima and Honda reported the ability of an illuminated TiO_2 and Pt electrode to generate the H_2 gas by a publication in *Nature*, which opened the doors for the close associations between photo electrochemistry and photocatalysis. By the end of that decade and by the mid of 1980s many other semiconductors, the ideas for nano-sized TiO_2 particles and coupled semiconductors had been evolved. The decades of 1980–2000 were important for the field of photoelectrochemistry as major breakthroughs during this era led to the discovery of large number of applications varying from self-cleaning surfaces to disinfection of water [2, 6, 7].

1.2 Semiconductor photocatalysis

The phenomenon gained importance during late 1970s and mid 1980s after the evolution of H_2 by illuminated TiO_2 (pristine semiconductor) in the presence of noble metal electrode, after that great concern was shown in water splitting followed by emergence of new era of photo electrochemistry at a single semiconductor crystal. The processes of semiconductor photocatalysis for environmental concerns have gained fame during the last 3 decades [8, 9].

Semiconductor photocatalysis can be of two types (i) homogenous photocatalysis, (ii) heterogeneous photocatalysis, depending on the phase differences of reactants and catalysts, triggered the concept of advanced oxidation processes (AOPs) at the end of twentieth century. TiO_2 is the semiconductor which can act as both the hetero and homogenous photocatalyst photoreactions [10] (**Figure 1**).

Heterogeneous photocatalysis is a process which includes a large variety of reactions: reductions oxidations, dehydrogenation, transfer of charge carriers, bacterial inactivation, organic pollutant degradation, water detoxification, etc. under both UV and visible light irradiation [1, 11]. Having TiO_2 as a semiconductor, the general mechanism is illustrated in **Figure 2**.

The chemistry behind the process is explained as [6]:

$$e^- \ CB + h^+ \ VB \rightarrow \text{energy} \quad (1)$$

$$H_2O + h^+ \ VB \rightarrow \bullet OH + H^+ \quad (2)$$

$$O_2 + e^- \ CB \rightarrow O_2^{\bullet -} \quad (3)$$

$$\bullet OH + \text{pollutant} \rightarrow\rightarrow\rightarrow H_2O + CO_2 \quad (4)$$

$$O_2^{\bullet -} + H^+ \rightarrow \bullet OOH \quad (5)$$

$$\bullet OOH + \bullet OOH \rightarrow H_2O_2 + O_2 \quad (6)$$

$$O_2^{\bullet -} + \text{pollutant} \rightarrow\rightarrow\rightarrow CO_2 + H_2O \quad (7)$$

$$\bullet OOH + \text{pollutant} \rightarrow CO_2 + H_2O \quad (8)$$

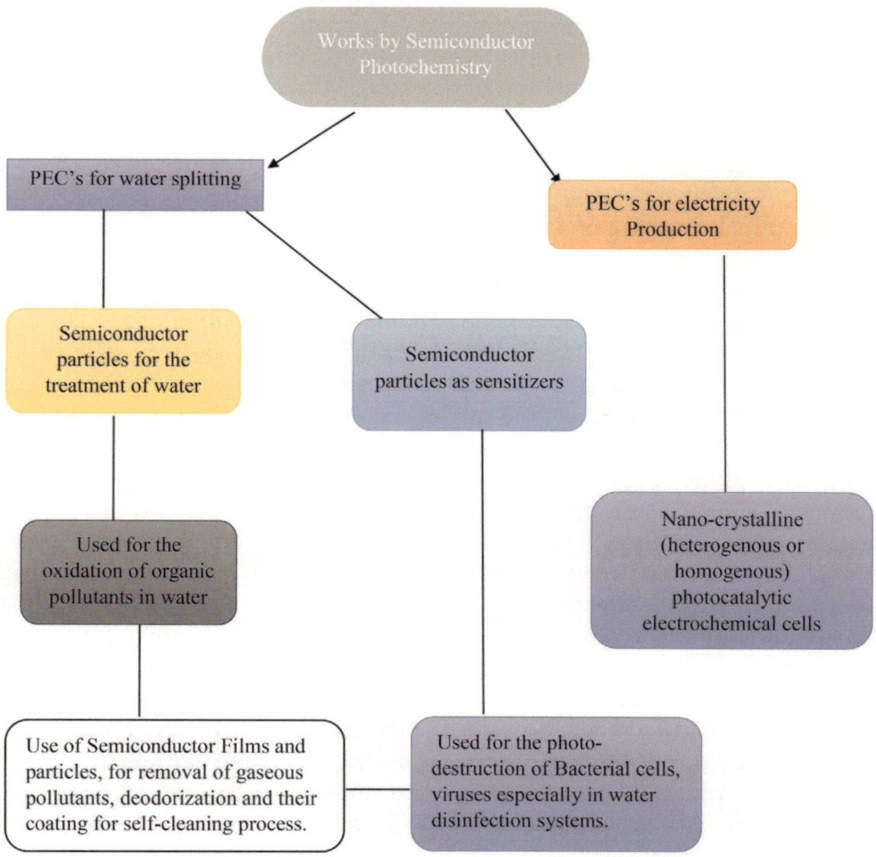

Figure 1.
Fields using semiconductors as photocatalysts from early ages to date [3].

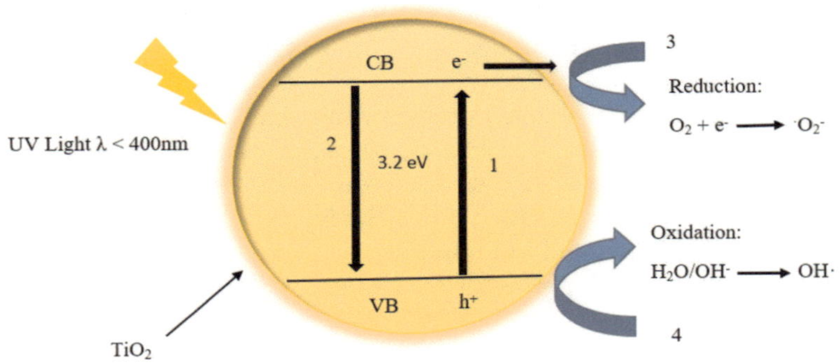

Figure 2.
Primary mechanism of photocatalysis of pristine semiconductor sphere illustrating different steps: (1) Formation of e^-/h^+ as a result of photoexcitation; (2) e^- recombination in absence of no electron acceptor; (3) photo induced electron transportation and reduction of oxygen; (4) oxidation due to hole in VB, while 3.2 eV is E_g between VB and CB.

1.3 Recombination

In **Figure 2**, process "2" is "recombination." When it occurs, the excited electron return to valence band and recombines with hole (by dispersing the absorbed energy as heat) in the absence of electron acceptor in the conduction band. Recombination of photoexcited electron is the major drawback of semiconductors as it lessens the efficiency of overall process. The chances of recombination can be reduced by modifications in the crystalline structure either by doping with metals, non-metals, ions or by the formation of heterostructures [12].

1.4 Basics of heterojunctions

In the internal structure of semiconductors (SCs), there are some band alignments which act as basis for the formation of heterojunctions/heterostructures; prepared for the enhanced activity of SCs' under sunlight. "Heterojunction is the junction in single crystal, formed by the combination of two dissimilar SCs." These heterostructures are formed on the basis of band gap, electron affinity and bands position thus leading to the formation of new energy levels. By the formation of HSs/HJs, the movement of charge carriers in a reaction can be controlled. There are three types of heterostructures depending on the energy band alignments. (i) Type I band alignment; the "Straddling" band Pattern. (ii) Type II band alignment; the "Staggered" band Pattern; (iii) Type III band alignment; the "Broken gap" Pattern. Agrawal [13] **Type I**, the most common band alignment in which the SC with smaller band gap relies under the SC with larger band gap; means band gaps are (not entirely but) overlapped. The charge carriers stay confined in the heterojunction due to the presence of potential barrier. **Type II**, the position of Valence and Conduction band of first SC is higher than that of second SC, the direction of steps in both SCs is same. The potential difference between two SCs causes the band bending at junction site due to this the electron and hole move in opposite directions, thus leading to good separation of charge carriers in heterojunction structure. When the band gaps stops to overlap then the case is known as "zero gap or broken gap" this is **Type III** band alignment, which is formed by the combination of semimetal having non-overlapping bandgap with a semiconductor [14, 15]. The illustration of band gap alignment is given **Figure 3**.

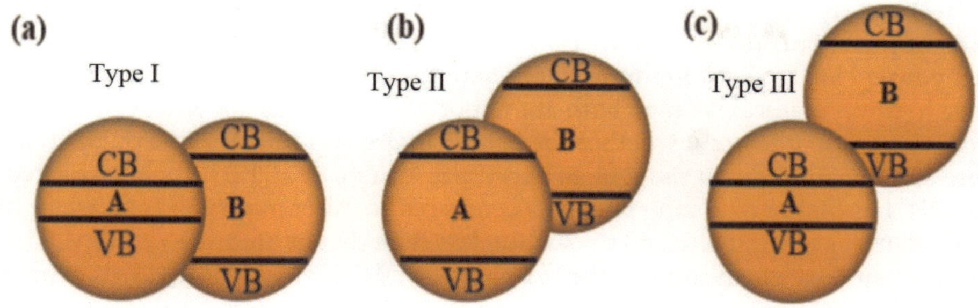

Figure 3.
Schematic illustration of different types of semiconductors heterojunctions [16].

The formation of Type II heterojunctions are good candidates in the field of photocatalysis as it has capability to enhance the charge separation to improve the photoreactions, photocatalytic degradation and water splitting [17].

1.5 General mechanism of water splitting by semiconductors for wastewater treatment

Globally, 1 billion people do not have access to safe drinking water and approximately 2.6 billion people are suffering from safe sanitation, this is particularly an issue in Asia, Africa, South and central America [18], so there is no ambiguity in saying that water management has become an emerging issue for the twenty-first century. Number of anthropogenic activities like industrial activities, discharge of effluents in the water bodies, oil spills and leaching of compounds to groundwater, etc. are interfering the quality of water, thus making it unfit for drinking and domestic purposes and also threatening the natural water resources. It has been estimated that the water quality will be a worst in upcoming decades. In this regard, there is a need for advancement in the fields of water and wastewater treatment because the conventional treatment plants are not capable to remove some persistent pollutants from water [19].

In this perspective the concept of photocatalytic water splitting has been emerged in last decade which are a favorable method for attaining green and renewable energy. For effective water splitting the CB of a SC should be more negative than the redox potential of H^+/H_2 (0 V vs. NHE) and the top of the valence band should be more positive than the redox potential of O_2/H_2O (1.23 V vs. NHE). In splitting process water breaks in oxygen and hydrogen by the photoexcitation of electron leaving a hole behind. Semiconductor absorbs the energy equal to or greater than its band gap, the photogenerated electrons and holes starts the redox reactions. Three key stages of water splitting mechanism are; (i) energy absorbance and generation of e^-/h^+ as result of photoexcitation; (ii) the charges remain separated at the opposite sides of SC crystal as a result of minimal recombination; (iii) charge carriers trigger the oxidation and reduction at valance and conduction band or vice versa. By production of reactive species as result of reduction of oxygen and oxidation of water or hydroxyl ion, like H^+, $OH^\bullet$ and $O_2^{\bullet-}$ the wide variety of organic pollutants in water are mineralized in to CO_2 and H_2O [20]. The water splitting mechanism is given in **Figure 4**.

1.6 Roles of reactive species in pollutant degradation

General reactive or oxidizing species formed during the photocatalytic degradation of pollutants are H^+, OH, HO_2, H_2O_2 and $O_2^{\bullet-}$. These photo-oxidants are formed by the steps as described in Eqs. (2)–(7).

Hole and hydroxyl radical: A hole generates when an electron moves to CB as a result of photo excitation, and hydroxyl radical generates when the hole oxidizes the hydroxyl ion or the water molecule. There is a debate on the reactiveness of hole and a hydroxyl radical, as they both are important for photocatalytic degradation, because it is still uncertain whether the hole or hydroxyl radical is core oxidant, or is there any dependence on the type of substrate. Minero et al. [21] has investigated that $OH^•$ has higher oxidizing power as compared to other oxidizing agents as it has the oxidation potential of +2.80 eV slightly less than that of Fluorine +2.87 eV. There are many studies which support the reactivity of $OH^•$ as an oxidant [22]. Turchi and Ollis [23] have proposed the four ways by which the $OH^•$ oxidizes the pollutant, **Figure 5**.

Draper et al. [8] has investigated the direct oxidation by the hole, it can be said that the oxidation process in the valence band is started by the hole, by the oxidation of H_2O molecule and hydroxyl ion. It would be rational to say that both the hole and $OH^•$ deal with different classes of compounds; like the production of hydroxylated rings from the oxidation of aromatic compounds and the degradation of paracetamol (acetaminophen) on the TiO_2 is caused by $OH^•$ [24].

Superoxide radical: This specie has been reported to play an important for the degradation/oxidation of many pollutants under both, the visible and ultra violet

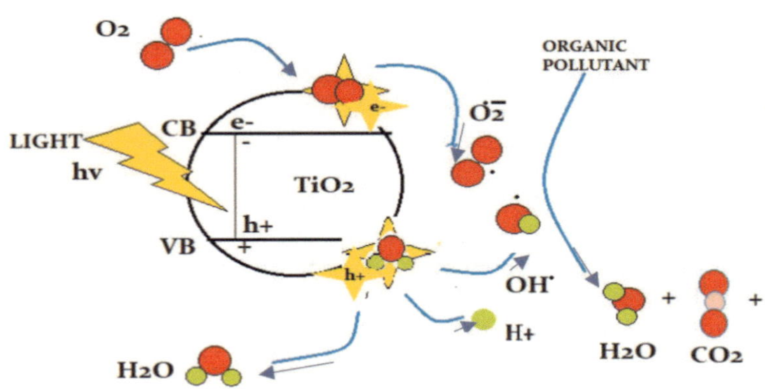

Figure 4.
Water splitting mechanism illustration and mineralization of organic pollutant.

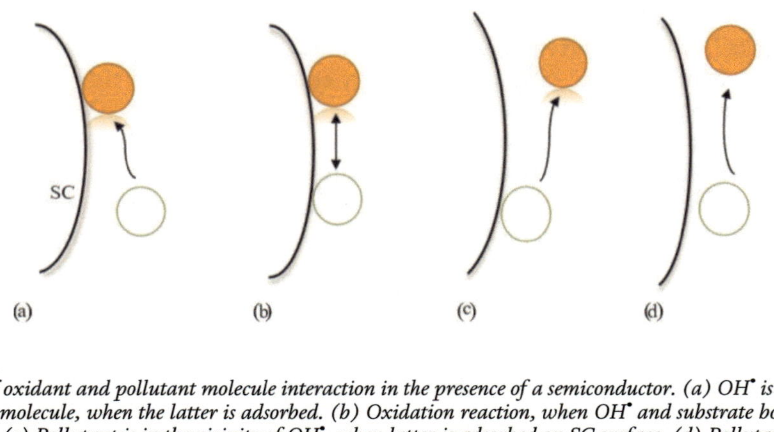

Figure 5.
Scheme of oxidant and pollutant molecule interaction in the presence of a semiconductor. (a) $OH^•$ is close to the pollutant molecule, when the latter is adsorbed. (b) Oxidation reaction, when $OH^•$ and substrate both are adsorbed. (c) Pollutant is in the vicinity of $OH^•$, when latter is adsorbed on SC surface. (d) Pollutant degradation by oxidation, when both are in solution; where ● is the target molecule/substrate and ○ is the hydroxyl radical.

light irradiations. In addition to direct participation in oxidation the formation of superoxides confirms the decreased recombination rate, as it is formed by the reduction of oxygen molecule. Therefore, superoxide formation is a vital process which controls the reaction rate by accepting the excited electron ([25]).

Hydrogen peroxide: Hydrogen peroxide is formed in the solution by the combination of two hydroperoxyl radicals (HOO·) or by the two electron reduction of O_2 in the conduction band as shown in Eqs. (6), (7) and (10). It can affect the photocatalytic reaction by acting as electron acceptor directly from organic or inorganic pollutant or by its dissociation into $OH^{\bullet}$ due to "homolytic scission" [8]. Reactivity depends on rate of its production and the substrate concentration on the semiconductor. H_2O_2 mostly in the TiO_2 solutions cannot be readily detected due to its high unstable nature. H_2O_2 is dissipated as it is produced into $OH^{\bullet}$ by reduction [21].

$$O_2 + 2e^-{}_{cb} + H^+ \rightarrow H_2O_2 \tag{9}$$

$$2H_2O + 2h_{vb}{}^+ \rightarrow H_2O_2 + 2H^+ \tag{10}$$

Photoproduced reactive species on the photocatalysts undergo oxidation that have been extensively studied during the last decades for environmental remediation including water disinfection, wastewater treatment, air decontamination, self-cleaning glass/surfaces degradation of organic compounds, etc. The reason to attraction towards this process lies in the end products of any type of pollutant which are CO_2 and H_2O. Actually, the mineralisation of compounds having high carbon hydrogen content, occurs via the formation of many intermediates which also undergo oxidation. In short, the significance of photocatalytic process and photo-oxidation cannot be underestimated because of its emerging demand for the decomposition of refractory organics or recalcitrant compounds.

2. Development of visible light active photocatalysts

Among the several advanced oxidation processes, semiconductor assisted photoreactions are gaining importance in recent years due to their higher mineralization potential of organic pollutants in environment. The main limitation for the pure SC photocatalysts is their large band gap which means there is a need for shorter wavelength of light and high energy photons ($\lambda < 380$ nm) to excite the electron from VB to CB. For this purpose, the photons falling in ultra violet region are required. Under the UV irradiation, the electrons in the VB of SCs are excited and e^-/h^+ pair is formed. The problem is, the UV light makes only 4–5% of the solar spectrum while about 40–45% of the solar light falls in visible spectra. The need of UV light limits the photocatalytic activity due to its less availability [12, 20].

In recent years, Yongquan Quab [26] has pointed some critical requirements for a stable SC catalyst to yield the solar energy which are; firstly, the SC must have an appropriate band gap to produce robust electrons ($E_g > 1.3$ eV commonly >2 eV but <3 eV) and sufficient band gap to allow effective absorption of light under visible region. Secondly, there should be less chances for recombination, i.e., should have an efficient charge separation system; lastly there should be a process to protect the semiconductors from the direct electrochemical reactions to confirm the photoelectrochemical strength of the system.

A single pure semiconductor photocatalyst is unable to fulfill the all above mentioned requirements. Therefore, in order to enhance the photocatalytic activity of the SC, there is a need to enhance their visible light activity, which can either be done by enhancing the surface modifications, i.e., by increasing the surface area and porosity, or by chemicals modifications like doping of metals, non-metals, ions, non-metal co-doping, dye sensitization or by the formation of junctions unitary, binary and tertiary, i.e., semiconductor/semiconductor, semiconductor/metal, semiconductor/metal oxides or the semiconductors/nano-composites [20].

2.1 Doping/grafting of semiconductor photocatalysts

Among all semiconductors TiO_2 is considered as pristine, first-generation oldest one with many physical and structural properties. In addition to TiO_2 other *d*-orbital metal oxides like WO_3, ZnO, Fe_2O_3 are regarded as *n*-type semiconductors while the semiconductors other than metal oxides are CdS, ZnS, CdSe, ZnSe, CdTe, MoS, Sb_2S_3 among them some have smaller band gaps like CdS and some have higher band gap like ZnS. TiO_2 is found in the three crystalline forms named as anatase, rutile and brookite, being the anatase and rutile are more active with band gaps or 3.2 and 3.0 eV. TiO_2 has been extensively used and widely investigated due to its high photo activity, biodegradability, less-toxicity, low cost and high structural and chemical stability against photocorrosion process [6, 12].

In order to understand, doping can be defined as "when impurities are added to semiconconductors, the band structure is modified; this process is known as doping." During the doping process, the doped atoms can present the interstitial, substitution or defect factor in structure of SC. When a semiconductor is doped with an acceptor atom, it is converted into *p*-type SC, because the acceptor atom is reduced by accepting electrons from VB and increases the production of holes and vice versa. Once SC is doped with any specie, it is supposed to increase the absorption of light in the visible spectrum and the doping material will not disturb the structural or chemical integrity of the SC [22]. The doping positions of dopants are given in **Figure 6**.

2.2 Non-metal-doped semiconductors

For this type of doping to SCs the non-metals like N, C, S and F have been used. The production of visible light active photocatalysts starts with the doping of N to TiO_2 or ZnO lattice, due to its small size, low ionization energy, and high stability. Livraghi et al. [28] proposed that contrary to the concept that nitrogen species are reason for formation of VLA photocatalyst the nitrogen precursor during the doping process induces the oxygen vacancies which renders the SCs able to absorb visible light and to increase phocatalytic activity. In 1986, Sato discovered the addition of NH_4OH in the titania sol followed by the calcination of obtained product, the resulting products was visible light active [29]. After that Asahi et al. [30] first time explored the VL activity of N-doped TiO_2 by the sputter method of TiO_2

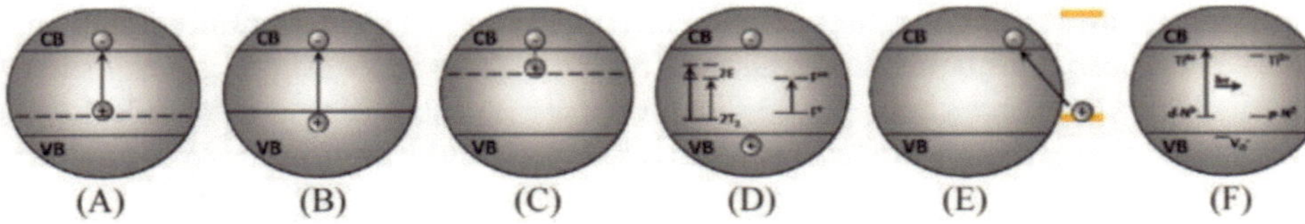

Figure 6.
Six schemes illustrating the different sites for dopants; (A) the localized states above VB; (B) lessened E_g due to non-metal doping; (C) localized states below CB; (D) formation of colored centers between E_g; (E) surface modification by addition of N-containing compounds; (F) interstitial N species and oxygen vacancies [27].

under N-Ar atmosphere. They also proposed that nitrogen doping creates a delocalized mixing/hybridization of O 2p and N 2p orbitals causing the rise in valence band position. Later on, for the efficient doping of N to TiO_2 either in bulk or at surface both types the dry or wet methods have been adopted. Techniques, like sputtering [31, 32] and ion implementation are based on the direct treatment of TiO_2 with nitrogen ions [32, 33].

The most used method of N doping in TiO_2 is sol-gel method, in this method the titanium precursors are combined with the nitrogen containing surfactants, for example, the N-doped TiO_2 have been prepared by using dodecylammonium chloride (DDCA) as surfactant acting as nitrogen source which provided it damaging capability for microcystins-LR under visible irradiation [34] **Figure** 7. Degradation of MC-LR and mineralization of 78% of carbon has also been reported by using Bi-doped TiO_2 under visible light irradiation having $OH^•$ as a reactive oxidation specie [35].

Several processes have been proposed to prepare the nanobelts by doping the 1D titania nanostructure with nitrogen following hydrothermal method having the anatase TiO_2 particles and NaOH as precursors and heating treatment with NH_3 ([25]). Production of N-doped TiO_2 nanotubes by the anodization of Ti in HF/H_2SO_4 electrolyte have been reported [36].

In addition to the N-doped TiO_2, ZnO can also be doped which is one of the most studied SC other than TiO_2 due to its applications in disinfection process and diversity of shapes but it has low stability. ZnO has also three crystalline forms, named as Zinc blend, Rocksalt and Wurtzite, having third one as most stable and active. The example of N-doped ZnO is preparation of zinc nanobundles by thermal treatment of (already prepared) ZnOHF nanobundles with NH_3 at different temperatures, as the temperature was increased the ZnOHF nanobundles were converted to N-ZnO nanobundles (**Figure 8**). These N-ZnO nanobundles gave dramatic increase absorption of light in visible region at $\lambda > 420$ nm [37].

Formation of the N-doped ZnO mesoporous nanospheres is an example of use of solvothermal treatment of $Zn(NO_3)_2.6H_2O$ which is the source of both Zn and nitrogen, in the presence of oleic acid, oleylamine, and octadecene. Mixture was heated at 260°C for 20 min and cooled at room temperature for 2 h followed by centrifugation, washing of precipitate and calcination at 400°C for 2 h led to the production of mesoporous nanospheres having size between 100–300 nm. This structure had the higher photoactivity as compared to pristine ZnO ([38]).

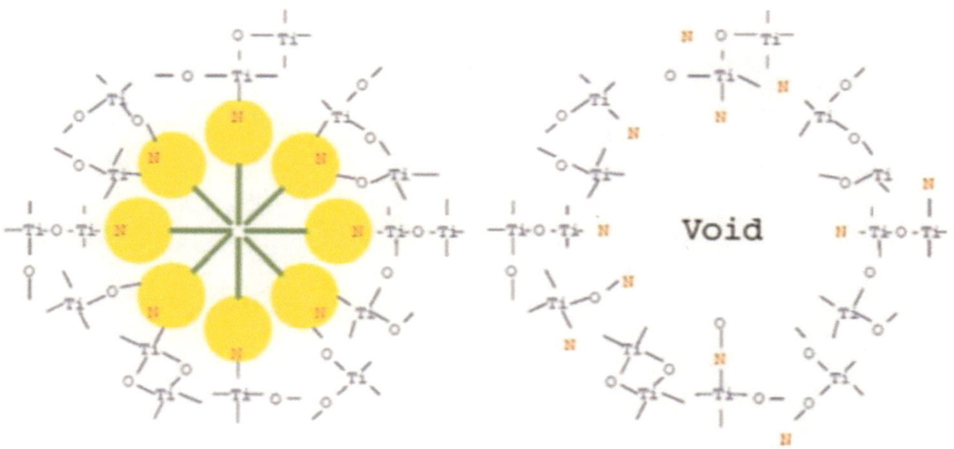

Figure 7.
N-doped TiO_2 using the titanium precursor and nitrogen containing surfactant as N source and pore template.

The nitrogen doped silver deposited ZnO thin films have been successfully prepared on a glass substrate by Kumar et al. via Radio frequency (RF) magnetron sputtering for the degradation of 2-CP under the visible light irradiation λ 390–700 nm and resulted in production of mineralized products as in **Figures 9** and **10** [39].

Nitrogen doped ZnO films were prepared by chemical vapour deposition (CVD) process using metallic zinc, NO_2 and NH_3 as precursors. The nitrogen concentration was varied while the temperature was kept constant at 350°C. The gradual addition of dopant to ZnO films led to the formation of stable nitrogen vacancies clusters and n-type defects, the stability of vacancies is the beauty of this process because the vacancies formed by addition of other impurities are easily removable [40]. Other techniques for the production of N-ZnO films on glass substrate, like high vacuum plasma-assisted chemical vapour deposition (HVP-CVD) [41] and pulsed-filtered cathodic vacuum arc deposition (PFCVAD) [42] have been reported. In both techniques, N_2 or N_2O was used as doping agents and as the dopant quantity was

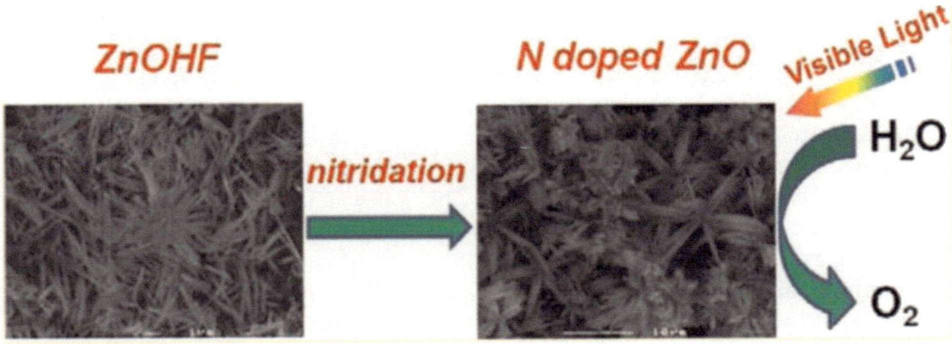

Figure 8.
Nitridation of ZnOHF nanobundles resulting in the formation of N-doped ZnO with similar morphology, showing high absorption to visible region.

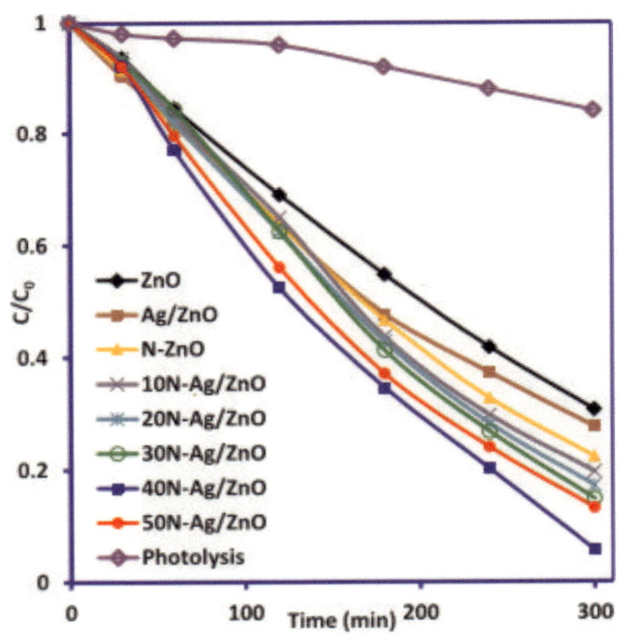

Figure 9.
Photocatalytic degradation of 2-CP on ZnO-based thin films under visible light irradiation [39].

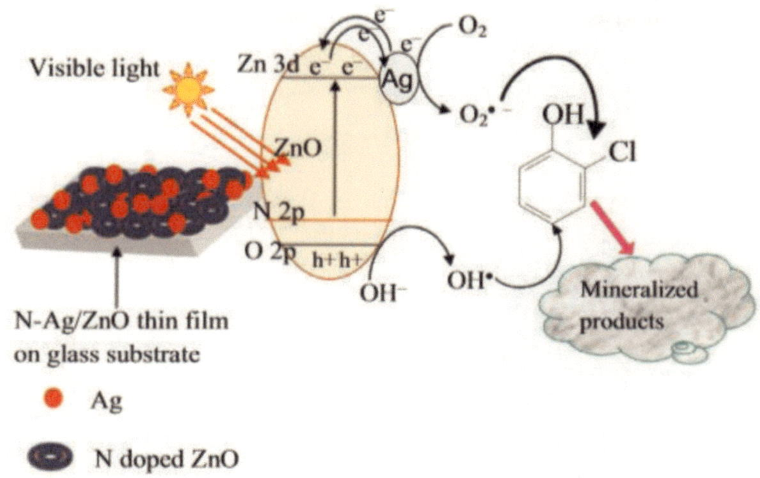

Figure 10.
Illustration of formation of e^-/h^+ pair and degradation of 2-CP by the N-Ag/ZnO thin films [39].

increased the semiconductor changed from *n*-type to *p*-type while the stability of *p*-type was dependent upon the synthesis process. Surprisingly the films synthesized by the PFCVAD process maintained their *p*-type doping for 12 months.

In literature, in addition to the doping of TiO_2 with nitrogen, there are studies which have reported the carbon doped TiO_2 as visible light photocatalyst. In contrast to the nitrogen doped TiO_2 which is commonly present in substitutional form, the carbon is doped to TiO_2 in different ways depending on the doping process, the three possible ways for carbon addition to lattice are (i) replacement of lattice oxygen with carbon; (ii) replacement of Ti atom with carbon atom; (iii) addition of carbon at interstitial position. Due to relatively smaller size of C atom, third one can occur without creating too much strain in structure [43]. The green synthesis for the C-doped TiO_2 has been described by using Ti $(SO_4)_2$ and sucrose as the precursors for Ti and carbon. This hydrothermal method is adoptable due to its low cost, nontoxic and easiness to perform, the C-TiO_2 was prepared by hydrothermal method and post-thermal treatment was applied at different temperatures and this was found to be effective to promote the visible light activity of C-TiO_2 for the degradation of toluene [44]. C-doped TiO_2 material have been prepared by the oxidation of titanium carbide (TiC) coated multiwalled carbon nanotubes (MWCNTs). The MWCNTs were used as a reaction template and carbon source, and titanium powder as the titanium source. XPS results indicated that chemical structure of Ti in TiO_2 coated-MWCNTs was different from that of pristine anatase due to the formation of Ti-O-C strong bond and interaction between TiO_2-MWCNT (**Figure 11**). The photocatalytic activity of this material was tested for the degradation of methylene blue under visible light [45].

Carbon doped TiO_2 has been reported to improve its VL response and photocatalytic activity and has been shown to be more effective than nitrogen doping. C-doped TiO_2 nanomaterials were prepared by the oxidative annealing of titanium (IV) carbide, used as a precursor. XPS analysis revealed that the carbon in the sample was present as carbonate species. The nanopowders were used for the degradation of methylene blue in the aqueous solution and for anti-bactericidal activity of *E. coli* k-12 and 80% inactivation in 30 min was observed [46].

Sulfur can also be used as a promising dopant to enhance the visible light activity despite of its difficulty to doping due to its large ionic radius. Sulfur-doped TiO_2 was prepared by using thiourea and titanium (IV) isopropoxide as precursors

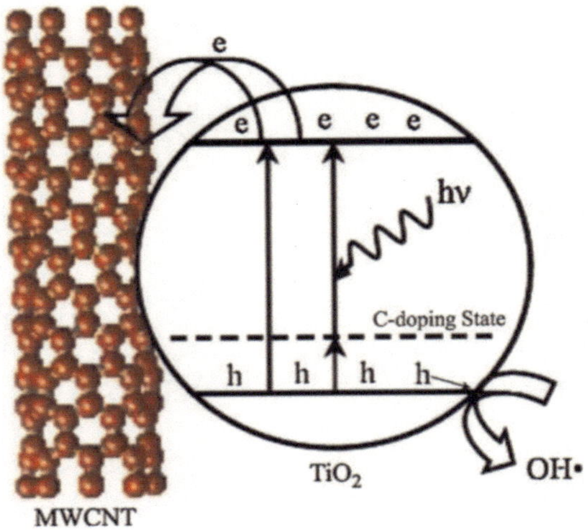

Figure 11.
Proposed mechanism of enhanced physical light activity of MWCNTs-TiO_2 material [45].

followed by mixing and vigorous stirring in ethanol resulted in the production of white powder annealed at 450°C for 4 h, the resulting material was S-doped TiO_2. These powders gave degradation of quinoline at λ = 495 nm [47]. S-doped material can be produced by the oxidation heating of TiS_2 powder followed by the annealing at 300°C, the XRD and XPS patterns gave the results which confirmed that the S atoms doped into the substitutional site of TiO_2 (e.g., the substitution of S for O) are effective for the band gap narrowing. Moreover, band calculations showed that the band gap narrowing due to the S doping originates from mixing the S 3p states with VB, leading to an increase in the VB width [48].

Incorporation of S to ZnO is little difficult due to its low solubility in ZnO and separation of ZnS particles at high temperatures. To overcome this problem, homogenously mixed precursors should be treated at low temperature. Formation of sulfur-doped ZnO films have been reported by pulse laser deposition [49], reactive sputtering ([50]) and chemical spray pyrolysis [51]. S-ZnO films are more appropriate for the uses in electronics, sensors and photoelectrochemical cells. But for the photocatalytic processes, S-ZnO powders give better results than films. Patil et al. [52] has proposed a method to make S-ZnO by a novel two steps process. First was the formation of bis-thiourea zinc oxalate (BTZO) powders by a mechanochemical method followed by its thermal decomposition/annealing at different temperatures to produce S-ZnO. At lower temperatures the resulting product was ZnS but as the temperature rose to 600°C the product was S-ZnO. The prepared powder was used to degrade the resorcinol under visible light and complete photocatalytic degradation (PCD) of 150 ppm resorcinol was observed in 7 h of light exposure at pH 7 [52]. ZnO has also been reported to be doped with iodine by following a solvothermal method using zinc salts and iodic acids in polyol medium as precursors with heating at 160°C. XRD and EDX analysis proved the presence of I in ZnO lattice. I doping increased the visible light absorbance, i.e., $\lambda > 510$ nm. In addition to the change in absorption spectra, it was proposed that doped I could act as the electron trapping sites thus reducing the chances for recombination [53].

To enhance the absorption in visible spectrum, P doping to TiO_2 has been reported using titanium isopropoxide and phosphide and H_3PO_4 as precursors. The content of P-TiO_2 had different properties prepared from two different precursors (Phosphide and H_3PO_4). A clear red shift in the absorption was observed due to the substitutional position of P^{3-} for oxygen in lattice. The presence of P^{3-} was confirmed by XPS analysis, the ion-doped product gave 4-CP and acetaldehyde degradation under pure visible light at λ 410–440 nm [54].

Graphitic carbon nitride (g-C_3N_4) is a polymeric semiconductor, in order to increase its visible light activity and to reduce band gap it is doping with phosphorous has been reported. Ligang et al. prepared P-doped g-C_3N_4 by one pot green synthesis approach using dicyandiamide and phosphorous containing ionic liquid as precursors. The prepared particles were characterized and analysed and were used for the degradation of organic dyes like Rhodamine B and Methyle Orange under visible light. The photoactivity of particles was dependent on the post annealing temperature, doped particles gave higher photocatalytic activity as compared to pure g-C_3N_4, and degraded 95% of pollutants for the irradiation time of 180 min [55]. Chai et al. has also reported the degradation of dye by using P-doped gCN under visible light, synthesized by co-pyrolsis method, analysis like XRD, SEM, FESEM, XPS and FTIR were performed to check crystalline, morphological, structural and optical properties of catalyst. Prepared catalyst gave greater photocatalytic activity by substituting P atoms with C resulted in enhanced light harvesting phenomenon as it degraded 95% of dye in 30 min, additionally radical scavenging test unveiled that holes and superoxide radicals were dominant reactive species [56]. Sulfur-doped gCN porous rods were prepared in one pot by a simple pyrolysis method of melamine-trithiocyanuric acid at different temperatures, analysis confirmed the formation of porous rods with high surface area than that of pure gCN, surface area increased with increasing temperature, as prepared catalyst gave higher photocatalytic activity and increased absorption of light in visible region and gave 92% of rhB degradation within 50 min [57]. g-C_3N_4 has also been reported to be doped with boron [58] and carbon atoms [59], to enhance the photocatalytic activity under visible light.

Some other non-metals like fluorine, bromine, and oxygen have also been reported to dope SCs to enhance the photocatalytic property. Doping of different type of dopant into the lattice structure of some other semiconductors like Ta_2O_5, Nb_2O_5, $BiVO_4$, $InVO_4$, Bi_2WO_6, $La_2Ti_2O_7$, $H_2Ti_4O_9$, $NaTaO_3$, Nb_2O_5,V_2O_5, Sb_2O_3, Bi_2O_3, Fe_2O_3, NiO, ZrO_2, CeO_2, Ga_2O_3, CuO, Cu_2O, HNb_3O_8, WO_3, ZnO, $K_2La_2Ti_3O_{10}$, $K_2Ti_4O_9$, $BiMoO_6$ and TiO_2 have been reported. Some doped SCs with their water treating abilities is given in **Table 1**. Among the above mentioned semiconductors some have the large band gaps and show excitation under the UV irradiation only, like ZrO_2 has the band gap of 5 eV and Ga_2O_3 have band gap of 4.8 eV needs the excitation energy of 248–265 nm, while some SCs like WO_3 and venedates have small band gaps but they are doped in order to enhance the quantum efficiency of the process. However, dependant on their synthetic approaches and crystalline structures some metal oxides have shown the degradation activity of organic and inorganic compounds under visible light [95].

2.3 Grafting of semiconductors by co-doping

2.3.1 Non-metal co-doping

Photocatalysts are not only doped with single **non-metal**, they can also be doped with more than one or two non-metals resulting to co-doping. A lot of work has been reported on this phenomenon in order to enhance the photocatalytic behaviour of SC. For example, N-C co-doped TiO_2 has been reported, being P25 as titanium dioxide source, ammonia as a nitrogen source and different types of alcohols as carbon precursors by using one step hydrothermal method. The mixture was placed at 100°C for 4 h. After thermal treatment, powder was cooled and dried for 24 h at 105°C, the obtained product was N-C doped TiO_2. The physicochemical

Pristine photocatalyst	Non-metal dopant	Doping process	Target pollutant for visible light induced degradation	References
TiO_2	Nitrogen	Colloidal solution hydrolysis	Methylene blue	[60]
		DC sputtering and oxidation	Rhodamine B	[61]
		Hydrothermal treatment	2-propanol	[62]
		Sol-gel, hydrothermal and pyrolysis	Microcystin-LR	[63]
		Thermal decomposition, one-step synthesis	MO degradation	[64]
		Microemulsion-hydrothermal method	2,4-dichlorophenol	[65]
	N-P	Sol-gel method	4-Chlorophenol	[66]
	Boron (F-B-S tri-doped)	Solvation/ evaporation/sol-gel synthesis	Acid napthol red (ANR)	[67]
	Boron		Trichlorophenol	[68]
	C-S	Solvation/evaporation	Toluene, 2-methlypyridine and MO, *E. coli* sterilization	[69–71]
	F-N	Hydrothermal/sol-gel/ Impregnation	MC-LR 4-CP MB	[72–74]
	C-N-S	Ti_2CN calcination/sol-gel	Reactive brilliant red X-3B, Tetracycline	[75, 76]
	S-C	Mechanochemistry	2-Propanol	[77]
$SrTiO_3$	N	Solvothermal/Sol-gel	MO, MB, RhB	[78]
$NaTaO_3$	I	Hydrothermal	MB	[79]
	N	Hydrothermal	Formaldehyde	[80]
ZrW_2O_8	S	Hydrothermal/ Calcination	O_2 Evolution	[81]
	I	Hydrothermal	RhB	[82, 83]
Bi_2WO_6	F	Hydrothermal HF	MB	[84]
ZnO	N	Decomposition reaction of Zn nitrate	MO	[85]
HNb_3O_8-SiO_2	N	SSR urea	RhB	[86]
WO_3	N	Thermal decomposition in NH_3	MeOH Oxidation, MO oxidation	[87, 88]
	F	Hydrothermal	rhB degradation Phenol	[89, 90]
$BiVO_4$	S	Aqueous thiourea	MB	[91]
$LaCoO_3$	C	Sol-gel process	CO_2 reduction	[92]
$In_2Ga_2ZnO_7$	N	SSR	H_2 Production	[93]
Ce_3O_4	F	PE-CVD	H_2 production	[94]

Table 1.
Visible light active non-metal doped photocatalysts.

properties and photocatalytic activity of catalyst was tested by using phenol as a model contaminant, phenol decomposition by different prepared materials confirmed that activity of catalyst was increased with the chain length of the alcohol precursor, and photoactivity of material was also evaluated [96]. A unique structured photocatalyst, C-S doped TiO_2 (TCS) was prepared by the hydrolysis of tetrabutyl titanate in a mixed aqueous solution of thiourea and urea followed by the stirring of solution at 12–24 h and water in mixture was dried at 80°C. Obtained precipitate was calcined at different temperatures, the obtained product was TCS. XPS analysis was performed which demonstrated Ti^{4+} ions were replaced by S^{4+} and carbonate ions. As prepared catalyst was used for the photodegradation of 4-chlororphenol under visible light (λ = 470 nm) [97]. Another example of C-S doped TiO_2 which is nanoporous highly reactive catalyst as TiO_2(C-TiO_2) and TiO_2 (S-TiO_2) was prepared at room temperature without heat treatment by using sol-gel method. Under visible light irradiation doped material showed high degradation of dye about 99% in 70 min. The catalyst also showed the antibacterial activity about 95% of *E. coli* was killed within 180 min even after 10 cycles for use of S-TiO_2 [98].

In addition to above mentioned non-metals, some halogens have also been reported for doping of TiO_2 [99, 100]. An example for halogens doping is preparation of Cl-B co-doped nanocrystalline titanium dioxide by a hydrothermal method using $TiCl_4$ as a titanium source and mixed hydrobromic acid (HBr) with ethanol as a bromine source, the doping of Cl^- and Br^- indicated narrowing of band gap and enhanced the photocatalytic activity for the overall water splitting process under visible light [101].

Microcystins (MCs) are the effective toxins which are produced and released by algae in the fresh and brackish waters upon cell rapture. Among dozens of congeners MC-LR is becoming a cause for disturbance to water quality and environment due to its concentration in water and high solubility and stability in water. Among many practices, N-F co-doped TiO_2 has been reported to degrade it by the attack of $OH^\bullet$ on the toxic ADDA (3-Amino-9-methoxy-2,6,8-trimethyl-10-phenyldeca-4,6-dienoic acid) chain under the visible light, normally the destruction of ADDA chain eliminates the associated toxicity but this breakage requires UV light <280 nm, but in case of N-F co-doped TiO_2 the production of free radicals due to photoexcitation, reductions and oxidations have reported to attack the ADDA and Mdha (Methyl-dehydroalanine group) chains and have resulted in the degradation of toxin and production of non-toxic intermediates and products [102].

Among titanates, co-doped K_2TiO_4, $SrTiO_3$, $La_2Ti_2O_7$, $K_2Ti_4O_9$ have been reported. The Co-doping of $SrTiO_3$ with carbon and sulfur has been reported. Co-doped product was produced by the calcination of mixture of thiourea and $SrTiO_3$ powders in lidded crucible at different temperatures. The addition of C and Si were confirmed by FTIR, XRD, XPS analysis, the obtained powder had the enhanced absorption shift from 400 to 700 nm. Cationic sulfur and tetravalent carbon doping increased the oxidative degradation of 2-propanol under visible light at λ 440 nm [77].

Nisar et al. [103] investigated the increased efficiency of $BiTaO_4$ by non-metal mono and co-doping in order to enhance its visible light activity. This SC has the band gap of 2.75 eV but it has ability to absorb only 19% of the visible light. Nitrogen doping to lattice created an electron acceptor state just above the VB in order to increase VBM. The lattice was co-doped with carbon and sulfur and after doping it was compared with the bulk. It was analysed that C-S doping led to the notable band gap reduction about 39% which was very close to the required band gap for a good photocatalytic material. The doped material had good overall water splitting ability [103]. Recently, co-doping of CeO_2 with Carbon and nitrogen has been reported to enhance its visible light photocatalytic activity. Analysis and calculations revealed the addition of non-metals in the lattice. C-N co-doping shifts the fermi level at the bottom of conduction band by creating impurities in the lattice followed by the increased absorption of light under visible region. The surprising fact about this catalyst was its absorption intensity between 400 and 600 nm was very high as compared to other co-doped and N-doped CeO_2 and C-doped CeO_2 [104].

2.3.2 Non-metal-metal co-doping

To enhance the photocatalytic activity of semiconductors, co-doping with metal and non-metals has also been reported. In case of TiO_2, Bagwasi et al. has reported the synthesis, characterization and application of Bi-B-TiO_2 nanoparticles. The co-doped nanoparticles were prepared by sol-gel method. Results indicated the bismuth and boron were doped in TiO_2 lattice, as Bi substituted Ti as Bi^{3+} which reduced the rate of recombination and B was present as interstitial and substitutional B which enhanced the visible light absorption. Bi-B co-doped samples showed better activities for degradation of acid orange 7 (AO7) and 2,4-dichlorophenol under visible light (λ = 700 nm) irradiation [105].

TiO_2 is also co-doped with boron and nickel in the form of its oxide. As mentioned in **Figure 12**. The addition of boron at substitutional or interstitial position increases the response to visible spectra while the loading of Ni_2O_3 further enhanced its photocatalytic activity. TCP, 2,4-DCP and sodium benzoate were chosen as target pollutants. TCP was not only efficiently degraded under the visible light but also its was mineralized, moreover the degradation of other two pollutants was also same as of TCP [68].

Doping of $NaNbO_3$ powders with Cr-N and Mo-N has been studied by using hybrid density functional theory. Co-doping with two pairs of metal/non-metal was tested separately to check the enhanced visible light activity of doped $NaNbO_3$. Metal atoms (Mo or Cr) replaced the Nb from the lattice while the O was replaced by nearest Nitrogen atom. Analysis and calculations revealed that monodoping of different dopants and co doping of Cr-N are inappropriate for photocatalytic decomposition of water under visible light because the defects formed due to dopants are above the fermi level so they cannot act as charge trapping sites thus resulting to increased rate of recombination. While, the co-doping of Mo-N proved to be an appropriate dopant as it reduced the band gap by creating different energy levels. Whereas the band gap reduction is dependent on the concentration of dopant material. Mo-N doped $NaNbO_3$ have overall enhanced water splitting ability and a promising photocatalyst for pollutant degradation ([107]).

2.4 Metal doping

Doping in the semiconductors mostly become a reason to point defects which introduces the levels or states near, above or below the valence and conduction bands which are identified and calculated by different analysis. When the metal is incorporated in a lattice, it is doped in the form of metal ion which produces a band or energy level in the forbidden energy area thus rendering the photocatalyst able to absorb visible light irradiation. Additionally, doped metal in the form of metal ion can change the equilibrium concentration of charge carriers by acting as an electron acceptor. Thus, increasing the overall quantum efficiency of the photocatalytic process [108]. The doping of semiconductors with metals/metallic ions has also been investigated in recent decades using many procedures including sol-gel, hydrothermal, solvothermal processes, etc. (as discussed earlier). Photocatalysts can be doped with transition metals like Fe, Cr, Mn, W, Ru, Rh, low cost, non-noble

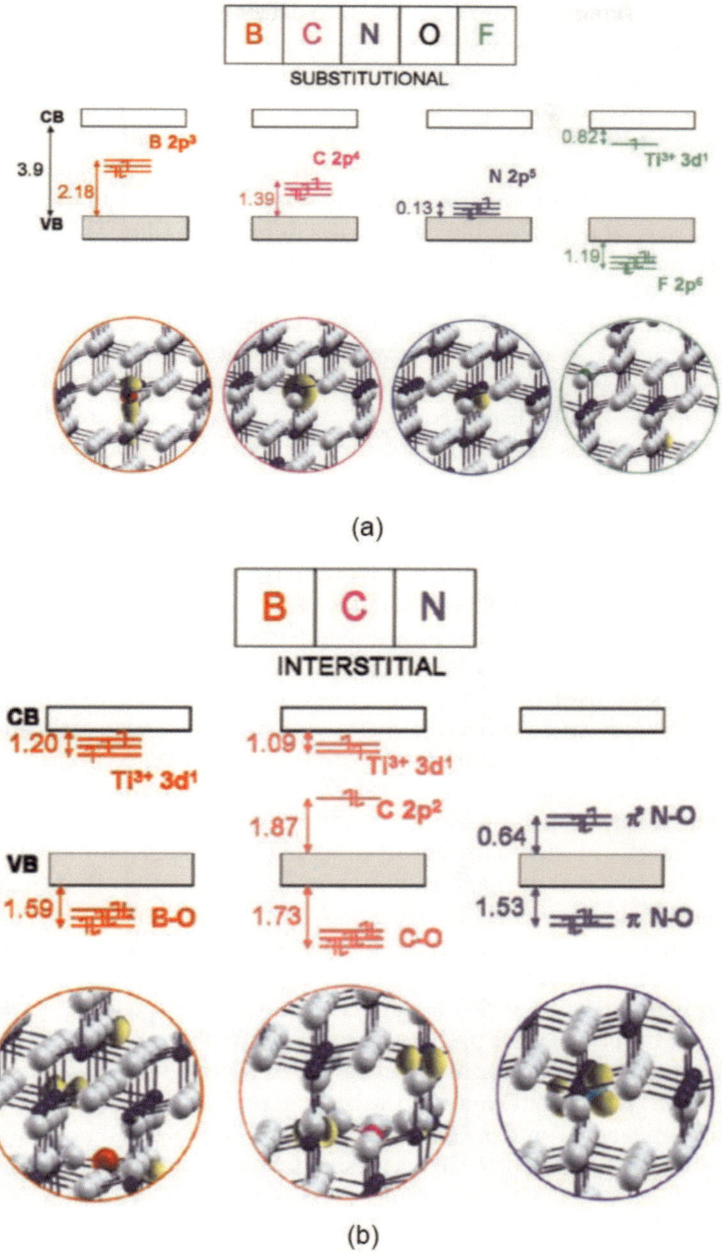

Figure 12.
Schematic illustration of positions of non-metals as dopants in the TiO_2 lattice. The effect of electronic structure of replacing O with B, C, N and F at substitutional and interstitial positions is considered; (a) doping of B, C and N atoms at substitutional positions results in magnetic impurities whose energy levels fall in the mid of energy gap. B sits high in the bandgap and N produces a state just above the VB while F being very electronegative occupies the state below O_2P of VB resulting the formation of TiO^{3+} ions; (b) at interstitial position, B behaves as a donor with the formation of B^{3+} and Ti^{3+}, carbon donates only two electrons and forma C^{2+}, nitrogen forms bond with O in lattice and does not donate electrons to host atom [106].

transition and earth abundant metals like Co, Cu, Ni and the noble metals like Pt, Au and Ag.

Preparation of **iron** doped TiO_2 thin films for the photocatalytic degradation of *Rhodamine B* has been reported by using reactive magnetron sputtering method having microscope slide as substrate and 99.99% pure TiO_2 as target and thin film was prepared by fixing pressure and flow rate. During the preparation of TiO_2 thin films, 99.9% iron pieces was placed on TiO_2 target to produce the doped plate the color of titanium dioxide film changed from light white to dark yellowish indicating the increasing concentration of iron. Analysis depicted as the concentration of Fe was increased, the wavelength of Fe-doped TiO_2 shifted to red due to excitation of Fe^{3+} electron to the conduction band. As prepared doped film was tested for the photocatalytic activity in comparison with undoped thin film and the degradation of rhB was observed under visible light on the doped thin film. The photocatalytic degradation rates of rhB was decreased surprisingly on the highly Fe-doped TiO_2 plates ([109]). **Vanadium** doped TiO_2 was prepared by sol-gel method with the objective to increase visible light response of TiO_2 having V^{4+} substituted at Ti^{4+} place in the lattice using vanadyl acetylacetonate, acetic acid and titanium butoxide as precursors followed by hydrolysis and calcination at 400°C. Analysis revealed that vanadium was highly dispersed inside the lattice and presented the red-shift, as prepared catalyst was used to degrade crystal violet (CV) and methylene blue (MB) under visible light irradiation which was higher than those of undoped powders [110].

Co-doping of N-doped TiO_2 with metals like **Sn** and **Zn** for enhanced activity under visible has been investigated and it is emerging as a promising approach to increase the photocatalytic ability of semiconductors because of synergistic effects in visible light absorption. Zn and Sn was compared for photocatalytic activity by doping with N (N + Zn and N + Sn), having tetracycline as a target pollutant. Different analysis (XRPD, HRTEM, XPS, EDX, and BET) for different calculations revealed that Sn and Zn modification directed to morphological, structural and surface changes, Sn was substituted at Ti site while Zn changed the lattice morphology. It was also determined that catalysts shown different performances in terms of light absorption, as Sn modified sample gave slight red shift by formation of intra-gap states and shown the absorption of light in range of 400–600 nm while Zn modification reduced the band gap by showing detrimental effect to crystallinity and creating surface defects [111].

Radio frequency (RF) sputtering technique was used to deposit **Ag** and ZnO nanoparticle on a substrate at different temperatures on a fixed pressure. Ag and ZnO having high purity of 99.999% targets were used as depositing materials and three different thin films were prepared depending on the substrate temperatures as shown in **Figure 13**. Characterization and analysis revealed the equal distribution of Ag nanoparticles within ZnO matrix, the band gap of the prepared nanoparticles was ranging between 2.7 and 3.1 eV while the pure ZnO has band gap of 3.37 eV. Photocatalytic degradation of 2-CP was tested by using prepared thin films by varying experimental parameters, results determined the enhanced photocatalytic activity of Ag/ZnO as compared to pure ZnO, reason for enhanced photocatalytic activity was the presence of Ag at the surface of film which acted as scavenger for an excited electron. The film prepared at substrate having highest temperature presented stability even after 4 cycles with only 8.7% efficiency loss [112]. Yayapao et al. prepared the undoped and Nd-doped ZnO nanoneedles using ultrasonic assisted solution method by varying the percentages of doped metal. Nd-ZnO nanoneedles gave the best results for the concentration of 1% Nd which was 50 nm in diameter and 3–4 μm long, gave performance about 2.5 times more as compared to the undoped ZnO. Analysis determined the presence of doped Nd^{3+} ions in lattice, which during photocatalysis acted as an electron scavenger and inhibited the process of recombination and promoted the photocatalytic activity by production of $O_2^{\bullet -}$ [113].

Undoped and **Ho**-doped ZnO 3D microstructures were prepared using a sonochemical method having zinc nitrate hexahydrate ($Zn(NO_3)_2{\cdot}6H_2O$), holmium

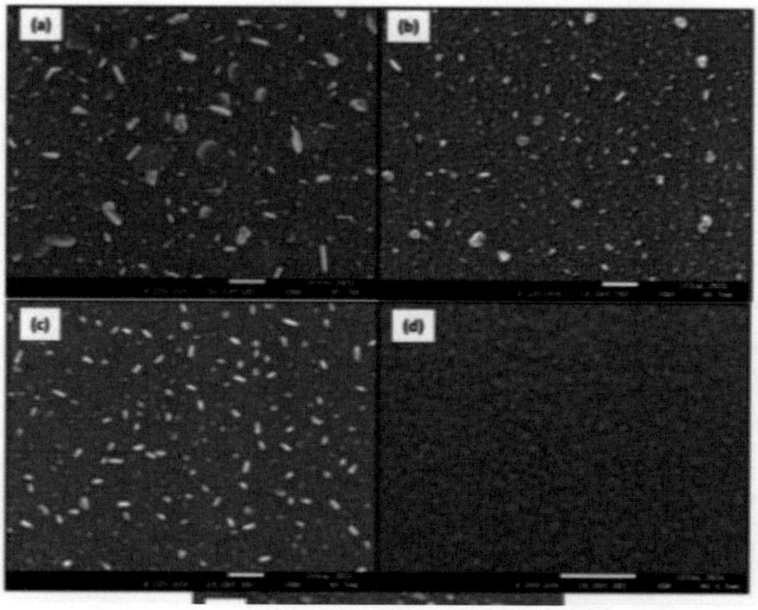

Figure 13.
SEM images of Ag/ZnO thin film co-sputtered at different temperatures (a) Room temperature, (b) 50 C, (c) 100 C and (d) Pure ZnO [112].

nitrate hexahydrate ($Ho(NO_3)_3{\cdot}6H_2O$) as precursors. The dopant concentration was varied from 0 to 3% after the sonication of 5 h the precipitate was washed and characterized. The flower like structure (**Figure 14**) gave the improved photocatalytic activity for the degradation of methylene blue (due to the production of non-selective oxidants) at λ 664 nm. The formation of Ho^{3+} ions at the dopant concentration of 3%, which acted as electron scavenger [114]. Similar positive results were reported by doping of La^{3+} and Sm^{3+} into ZnO lattice for the degradation of 4-nitrophenol in wastewater [115].

For **Copper** doped $BiVO_4$, the monoclinic crystal lattice ($mBiVO_4$) shown the efficient charge carrying capacity due to its smaller effective mass as compared to Ag and Au, which is promising for higher mobility of electrons and holes. In this system the substitution of Cu 3d states at the Bi sites reduces the band gap and also act as an electron acceptor produced by photoexcitation. This doped catalyst has shown the red shift which meant it enhanced the solar light absorbance and utilization, the system has overall great water splitting capacity [116].

Besides TiO_2 and ZnO, $\mathbf{WO_3}$ is also a promising semiconductor found to degrade the pollutants in water due to its photocatalytic activity with the band gap of 2.8 eV. Despite of short band gap, it can absorb only small portion of visible light. The photocatalytic activity of WO_3 was found to be amplified in the occurrence of dopant. WO_3 was doped with different metal ions using different precursors, each precursor was mixed vigorously with WO_3 to ensure the homogenous mixing followed by the drying and calcination for 4 h at 550°C, the final product was prepared after cooling and grounding of powder. Analysis confirmed the presence of different ions like Mg^{2+}, Al^{3+}, In^{3+}, Fe^{3+}, and Zr^{4+} on the interstitial positions in the WO_3 lattice. Doped SC gave significance change in the photocatalytic property in comparison to pristine WO_3 [117]. Palladium doped WO_3 was used to remove Geosmin (GSM) from the wastewater. Nanocatalyst was prepared by mechanical mixing in a ceramic mortar having Pd and WO_3 powder as a precursor, analysis revealed the size of Pd nanoparticles of about 10 nm which was dispersed evenly on the surface of WO_3, the photocatalytic degradation was done by spiking of GSM and catalyst in a solution with water (**Figure 15**). Oxidative species analysis depicted that $OH^{\bullet}$ was dominant oxidative specie. Degradation rate in initial 5 min was 75% indicating higher degradation rate. While in next 20 min the degradation rate was more than 99%, as the amount was catalyst increased gradually. The degradation rate decreased mainly due to the accumulation/agglomeration of catalyst and turbid environment caused hindrance to light penetration. The stability of Pd-WO_3 nanocatalyst was tested which retained 94% stability even after the fourth cycle which suggested its reusability for the GSM degradation [118].

Figure 14.
SEM image of holmium doped ZnO microstructures with (a) 0%, (b) 1%, (c) 2% and (d) 3% of HO having100–400 nm diameter and length of several micrometers [114].

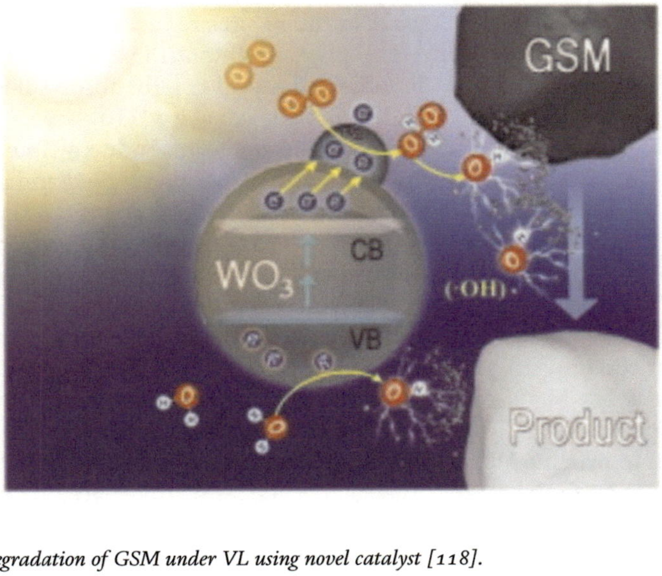

Figure 15.
Photocatalytic degradation of GSM under VL using novel catalyst [118].

Electronic, geometric and optical properties of **chromium** doped $SrTiO_3$ has been studied by Wei et al. by using DFT with GGA scheme. Results revealed that in Cr-doped $SrTiO_3$ structure the VBs did not suffer from any change while the bottom of CB shown a little rise. Cr substituted in lattice at the site of Sr. as Cr^{3+} while band structure and DOS analysis shown that there were also some Cr 3d gap states appeared near the bottom of CB. While Cr substituted at the site of Ti^{4+} was doped as Cr^{6+}. So the results indicated that doped chromium partially take up Sr. and partially Ti, Cr dopants at Cr-Sr site has higher photocatalytic activity as compared to the photoactivity at Cr-Ti dopants sites. Similarly the $SrTiO_3$ doped with less concentrate on of dopant shown less photocatalytic activity and less absorption of visible light and vice versa [119].

Dysprosium doped WO_3 nanopowders were prepared using $Na_2WO_4{\cdot}2H_2O$ and CTAB as precursors followed by a hydrothermal treatment at 80°C for 1 week. The resulting precipitate was centrifuged and calcined to obtain Dy-doped material. Analysis revealed that metal was doped as Dy^{3+} which acted as an electron donor to the adsorbed O_2 to make superoxide radical and Dy^{4+}, dopant ion also reduced recombination rate by accepting the exciting electron and formed Dy^{3+}, nanopowders showed high photocatalytic activity for the degradation of rohdamine B (rhB) which was used as target pollutant [120]. **Gallium**, a post transition metal has been shown to dope semiconductors. Li et al. have reported the increased charge separation and photocatalytic activity by Ga-doped SnO_2 in different molar ratios varying from 1 to 4%, using $SnCl_4$, $Ga(NO_3)_3$ and HNO_3 as precursors followed by drying and baking at 573 K for 2 h. The resulting product was Ga-doped SnO_2, analysis confirmed the presence of Ga as Ga^{3+} and shifted the absorption spectra towards blue region. To understand the charge separation, production of reactive species and photocatalytic efficiency methyl orange (MO) was used as a target pollutant, mainly due to the formation of $OH^{\bullet}$ MO was photocatalytically decolorized [121, 122].

Hu et al. reported the band gap-tunable K-doped g-C_3N_4 which had enhanced mineralization capacity, using dicyandiamide monomer and potassium hydrate as precursors followed by mixing, heating and annealing. SEM, TEM, FTIR, XRD and XPS analysis were used for the characterization and catalyst was used for the degradation of organic rhB under visible light and showed 6.5 times greater photocatalytic activity as compared to pristine gCN as it gave 70% decomposition during the irradiation time of 120 min [123]. Tungsten doped porous g-C_3N_4 was synthesized by hydrothermal method using urea, dicyandiamide and $Na_2WO_4.2H_2O$ as precursors and catalysts with varying molar ratios of precursors were prepared and the catalyst W5/PGCN (P stands for porous) gave the highest photocatalytic activity due to separation of charges and shown 99.6% degradation of MO under visible light (λ = 400–420 nm) for the irradiation time of 60 min and 100% in 70 min [124]. Beside W, other transition metals like Yttrium ([125]), Iron [126], Eu, Zr [17], etc. doped gCN for the increased photocatalytic activity has also been reported in literature.

3. Dye sensitized VLA semiconductors

Dye photosensitizations have been reported by different scientists being a most operative mode to increase the response of semiconductors for visible light. Sensitization has been used for different approaches, from development of solar cells to degradation of pollutants in water. Many dyes like erythrosine B, eosin, rose bengal, rhodamine B, cresyl violet, thionine, chlorophyllin, anthracene-9-carboxylic acid, porphyrins, phthalocyanines and carbocyanines have been reported for their work as sensitizers, in addition to these organic dyes some inorganic dyes and coordination metal complexes are also reported. Inorganic sensitizers are basically the semiconductors with smaller band gap which are highly stable to photodegradation or photocorrosion. Their smaller band gap gives the light absorption at the wider wavelength region and examples are Mn, Fe, Ni, V, and Cr [127]. The principle of sensitization for photodegradation of pollutants is "an electron in a sensitizer molecule under visible light (as dyes have conjugate system that absorbs visible light) is excited (from HOMO to LUMO) to its conduction band from where it jumps quickly to the conduction band of attached semiconductor oxide and dye itself changed into its cationic radical and SC's CB act as an electron mediator where it is accepted by oxygen, thus starting the oxidations reactions by forming superoxide and $OOH^{\bullet}$, which are then converted to $OH^{\bullet}$ for the degradation of pollutants"[12]. A main limitation/drawback of sensitization with an organic dye is its steady decomposition because of photocatalytic degradation due to its natural capacity to undergo redox reactions by donating its electrons, to dodge this limitation the addition of sacrificial electron donor is possible like the incorporation of EDTA and TEOA (triethanolamine) will improve the stability of dye. In a dye sensitized process the dye can be used as both a sensitizer or a substrate to be degraded in order to decolorize water [128]. The general mechanism for dye sensitized process for example rhB is given as:

$$rhB\ (ads) + hv\ (554\ nm)\ rhB* \left(\text{excitation of dye electron}\right) \tag{11}$$

$$rhB* + SC\ rhB^{\cdot +} + SC\ (e^{-}{}_{cb})\ (\text{electron 3 transfer to SC}) \tag{12}$$

$$SC\ (e^{-}{}_{cb}) + O_2\ O_2^{\cdot -} + \text{SC}\ (e^{-}{}_{cb})\ (\text{electron acceptance}) \tag{13}$$

$$\text{rhB}^{\cdot +} + \text{O}_2^{\cdot -}\ \text{CO}_2 + \text{H}_2\text{O}\ (\text{mineralization process}) \tag{14}$$

Many organic dyes have been tested for their excellent sensitizing ability to semiconductors like TiO_2. Degradation of phenylurea herbicide monuron has been reported by using riboflavin-assisted photosensitization, an enhanced photodegradation effect was observed as compared to direct photolysis [129], degradation of micosporine like amino acids (small secondary metabolites produced by organisms that live in environments with high volumes of sunlight, usually marine environments by some type of algae) was observed by using rose bengal and riboflavin sensitizers [130]. Degradation of dye named reactive red 198 has been reported under visible light irradiation by using dye sensitized TiO_2 activated by ultrasound, degradation rates were analysed by varying experimental parameters like initial concentration, pH and catalyst loading, the singlet oxygen and superoxide radical were found as dominant oxidative species which also degraded the produced intermediates like phenols. The results determined that a conventional dye can be used as a photosensitizer of TiO_2 functioning under visible light [131]. A study investigated the application of TiO_2 sensitized by tris(4,4′-dicarboxy-2,2′-bipyridyl) ruthenium(II) complex for CCl_4 deprivation under visible light irradiation. The transferred electrons to the CB of TiO_2 from the excited complex molecule decomposed CCl_4 at $\lambda > 420$ nm, experiment in the absence of TiO_2 revealed that there was no direct transfer between excited sensitizer complex molecule and pollutant, results also determined that sensitizing complex undergo redox reaction and photodegradation decreased in the absence of sacrificial electron donor which restore the Ru^{II} complex which undergo continuous oxidations by electron transfer, to avoid this problem different types of alcohols can be used but 2-propanol was used for $Ru^{II}L3/TiO_2/CCl_4$, in this process CCl_4 was decomposed in CCl_3 and $Cl^{\cdot}$ [132].

4. Heterostructures/heterojunctions/nano-composites

Construction of heterojunctions or nanocomposites is a method to increase the charge separation and to increase the absorption towards visible light region in the solar spectrum. As discussed above, the increased charge separation will reduce recombination resulting in increased production of reactive oxidizing species leading to enhanced degradation of contaminants in water and wastewater treatment. The crystal lattice structure at the point of interface or junction plays an important role in tailoring the photocatalytic properties or quantum efficiency of the photocatalyst. A difference in the lattice spacing of two SC crystals cause the lattice mismatch at junction point which is the reason for defects formation which act as excited electron sink and lessens the electron hole fusion [133]. Careful selection of semiconductor material is required with respect to band edge positions. The band edge positions of widely used semiconductors are given in **Figure 16.** The traits for ideal nano-heterostructures which have to be used for photocatalytic processes are; (i) it should have large surface area and contain enough active sites; (ii) it should have high light absorption capacity to make efficient use of solar spectrum; (iii) there should be an effective separation of charge carriers in order to produce active radicals for pollutant degradation; while a co-catalyst in heterostructure should have ability to enhance the reaction rate, provide active sites and should act as electron scavengers [134]. Due to their increasing demand for the degradation of pollutants controllable fabrication of heterostructures/nano-composites is possible, the different techniques which are being used over decades are chemical vapor deposition, chemical deposition, electrodeposition, etc. By using these techniques different types of heterostructures (capped and coupled) such as nano-composites, nano-rods, nano-sheets, nano-wires and core-shells (capped) structures have been reported. In comparison to single/alone SC the heterojunctions like CdS/TiO_2, CdS/ZnO, ZnSe/ZnO, ZnS/ZnO, SnO_2/CdS, Bi_2S_3/TiO_2 and some ternary composites like Fe_3O_4/SiO_2/TiO_2, ZnO/TiO_2/CuO, etc. have been reported.

The heterojunction which are made by the combination of two or more nanocystalline semiconductors are very fruitful for photocatalysis regarding pollutant degradation and water splitting and being very famous as the Z-scheme process. Two main benefits of linking two or more semiconductors are (i) coupling of semiconductor having larger band gap with the semiconductor having smaller one, amplifies the photo response acting as a sensitizer; (ii) recombination is avoided by the transfer of excited electron in the low lying CB of SC bearing larger band gap [135]. So, it is crucial to find a suitable sensitizer for a semiconductor having larger band gap to get enhanced photocatalytic activity under visible light.

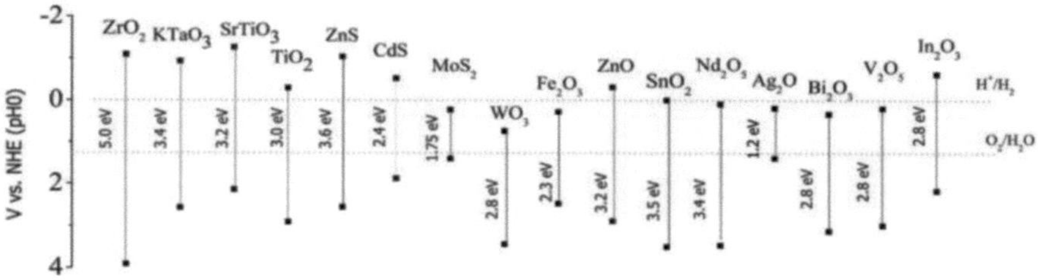

Figure 16.
Band gaps of different semiconductors with respect to NHE [17].

4.1 Some TiO_2-based heterostructures/nanocomposites

CdS is an attractive semiconductor with a smaller band gap of 2.4 eV, which means it can be activated under visible irradiation. The drawback to this and some other semiconductors having smaller band gaps is their unstability against photocorrosion. To avoid this problem, this type of semiconductors are coupled with those which have larger band gap like ZnO, ZrO_2, TiO_2, etc. and play a vital role in improving charge separation. As mentioned above there are many strategies to prepare the heterostructures and one is micro-emulsion method. CdS nanoparticles were coupled with TiO_2 via micro-emulsion method under ultrasonic irradiation for the formation of core-shell type of nanostructure with uniformed coating of TiO_2 nanoparticles on CdS, the depth of TiO_2 shell was in range from 1.4 to 2.3 nm which could be controlled by varying the preparation temperature and precursors concentration. The nanostructure gave the red-shift in spectra which is typical trait of core-shell heterojunction [136]. Another TiO_2/CdS core-shell nano-rod heterostructure was obtained by chemical bath deposition method, TiO_2 core was prepared firstly by the hydrothermal method and was converted to substrate for the deposition of CdS nanoparticles by dipping it an aqueous bath having 10 ml 0.02 M $CdCl_2.2.5H_2O$, 10 ml 0.2 M CH_4N_2S, 10 ml of 1.5 M NH_4NO_3 and 10 ml 0.5 M KOH to adsorb CdS nanoparticles on TiO_2 rod at 85°C. The resulting structure gave the highest charge separation (**Figure 17**) efficiency at incident wavelength of 500 nm [137].

A thin film heterostructure of copper zinc tin sulfide (Cu_2ZnSnS_4)-TiO_2 was prepared by combining an *n*-type (TiO_2) and *p*-type (CZTS) heterostructure. To confirm the formation of visible light active photocatalyst it is suggested that the highly stable semiconductor with higher band gap like TiO_2 should be present at most upper/outer layer in the heterostructure which will protect the

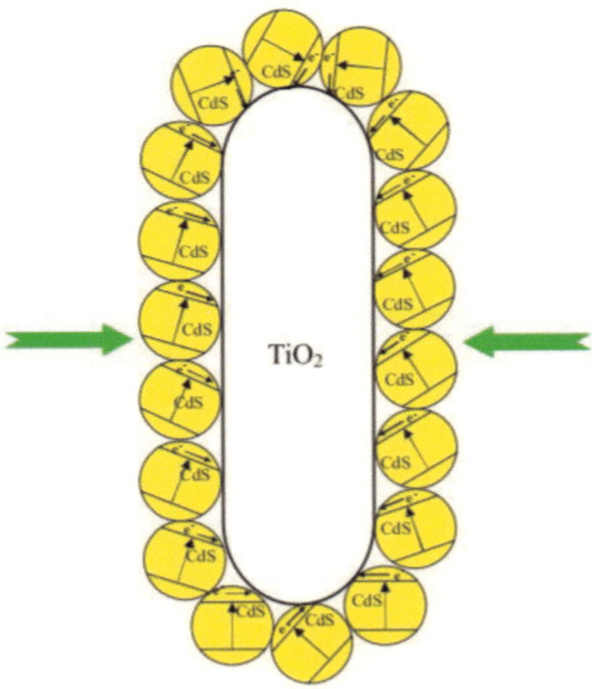

Figure 17.
Scheme illustrating charge injection from excited CdS nanoparticles to TiO_2 rod [137].

semiconductors having smaller band gap to come in direct contact with wastewater/water in order to avoid physical or chemical reactions and to absorb penetrated light. In the present study the thickness of outer layers of TiO_2 was varied by changing the concentration of precursor ($TiCl_4$ 0.1 and 0.05 M) and two types of catalysts also tested for photoactivity, uniform crystallinity depended on the thickness of the TiO_2 film. The schematic illustration of photodegradation mechanism is given in **Figure 18**. Analysis like TEM, SEM and EDX discovered the presence of multiple TiO_2 aggregates on the surface of CZTS thin films which have band gap of 1.46 eV. The catalyst prepared from 0.05 M solution exhibited poor Cu and enhanced Zn deposition which led to enhanced charge separation in the composite. Methylene blue dye was degraded under visible light by 0.1 M sample and shown efficiency of 43.50% while 0.05 M sample gave 35.7% removal under wavelength of 400–700 nm. In dark conditions TiO_2-CZTS of 0.05 M gave efficiency of 29.81% after 8 h [138].

Recently, the preparation of magnetic core-shell-shell heterostructure nanocomposite of $Fe_3O_4/SiO_2/TiO_2$ has been reported for the degradation of 2-CP in unreal wastewater under UV irradiation. Catalyst was prepared by three steps-facile synthesis process. Analysis was performed for surface morphology, chemical properties and for crystal structure. The size of NPs of magnetic core was 24.0 nm and that of magnetic shell-shell was 70.2 nm. Experiments were performed by varying the experimental parameters like pH, catalyst loading and pollutant concentration; results determined that effective degradation was pH dependant. Catalyst was regenerated after use by placing magnet under reaction container and retained its efficiency about 60% even after 3 cycles of use [139].

Another example of heterojunction is coupling of Ag_3PO_4 with TiO_2 to degrade the organic pollutants under visible light irradiation using silver nitrate, sodium phosphate and TiO_2 Degussa (P25) as precursors followed by in-situ precipitation method, by coupling the semiconductors the band gap reduced to 2.3–2.5 eV and shown a notable shift from UV region to visible spectra. Analysis confirmed the size of particles and uniform crystallinity 2-CP was selected as the target pollutant because this parent compounds and derivatives are being an emerging pollutant and

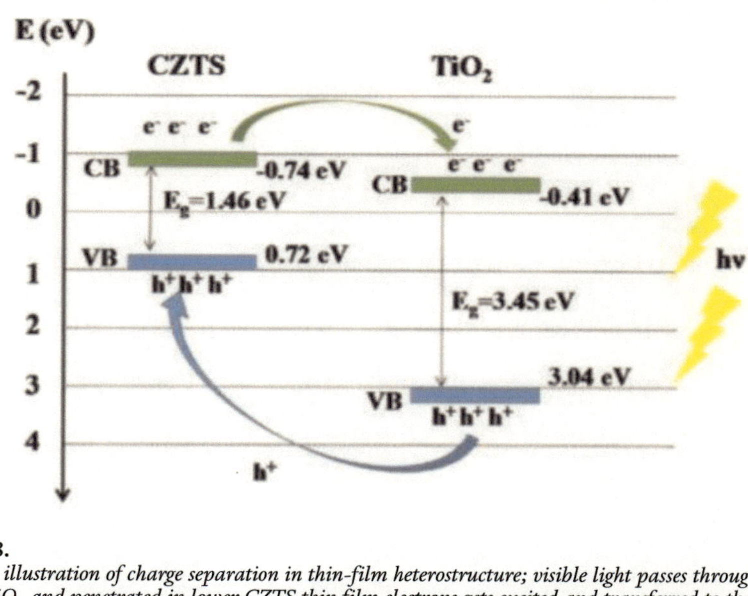

Figure 18.
Schematic illustration of charge separation in thin-film heterostructure; visible light passes through the upper layer of TiO_2 and penetrated in lower CZTS thin film electrons gets excited and transferred to the CB of TiO_2 creating and leaving the hole behind; electron in the VB of TiO_2 gets excited after exposure to UV light and hole produced diffuses to the VB of CZTS thus completing the circuit, while the holes at the VBs act as oxidants and electrons in CBs act as reductants to produce oxidizing species [138].

already in the POPs. Experiments were performed by changing the parameters like pH, catalyst loading and pollutant concentration. At low pH of 3 greater adsorption of pollutant was observed and maximum degradation was reported. The structure maintained its stability to 100% (**Figure 19**) even after 3 cycles of use with fresh 2-CP [112].

Pan et al. has reported a novel structure of TiO_2-ZnO nanocomposite spheres decorated with ZnO, this structure was prepared in order to increase the photocatalytic activity by improving charge separation in heterostructure. The composite was prepared by one-pot solvothermal method using tributyl titanate and zinc acetate as precursors and heating the mixed solution at 150°C for 24 h followed by calcination in air, as-prepared product was characterized. Surprisingly, by increasing to amount of Zn precursor the clusters of ZnO were produced on the surface of spheres (**Figure 20**). Analysis determined that prepared ZnO and TiO_2 had the particle size in spheres was about 4 nm and spheres' surface thickness was about 5 nm. It was also discovered that the clusters which decorated the spheres

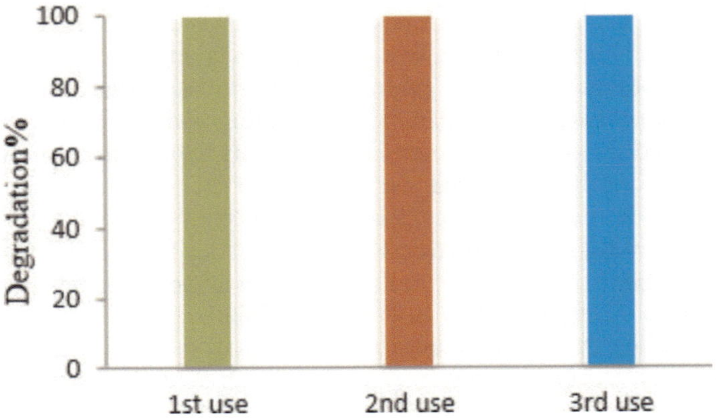

Figure 19.
Retained catalytic stability even after 3 cycles of use with fresh 2-CP [112].

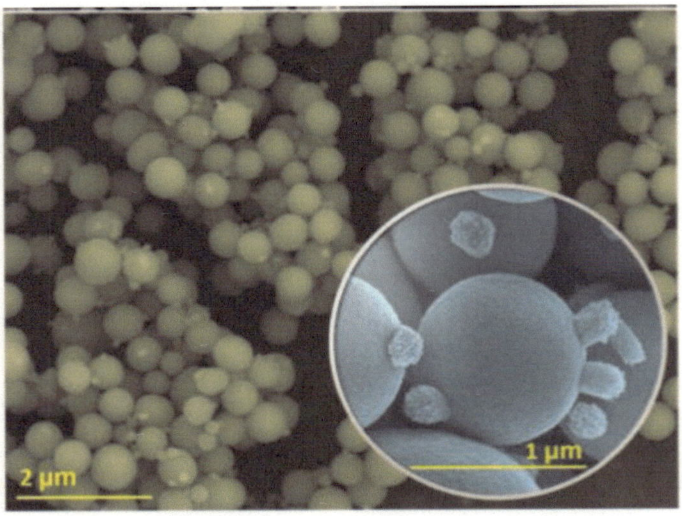

Figure 20.
SEM image of TiO_2-ZnO-SC, showing the decorated clusters of ZnO on the TiO_2-ZnO nanocomposite [140].

only had Zn and O and did not have Ti atoms. PL spectra for composite indicated the increased charge transfer and separation as compared to pristine ZnO and TiO_2 due to presence of ZnO clusters acting as electron scavengers. To investigate the photocatalytic performance of ZnO-TiO_2-SC rohdamine B (rhB) degradation was tested, $OH^{\bullet}$ were reported to be the reactive specie, the activity of ZnO-TiO_2-SC for photodegradation was 2.3–2.8 times higher than that of pristine nanoparticles [140].

Stable preparation of core/shell nanocomposite of TiO_2/FeS_2 by solvothermal method has been reported for the photocatalytic production of hydrogen [141]. The benefits for incorporating different metal oxides/sulfides led to the structural and chemical modifications which render them better than that of parent materials. Rashid et al. has reported the formation of stable spherical nanocrystals of anatase-TiO_2/FeS_2 by following wet chemical process, for the degradation of organic pollutants under visible light using $FeSO_4.7H_2O$, $Na_2S_2O_3$ and $TiCl_4$ as precursors. Micrographs and other analysis determined the morphology, chemical properties, crystal structure, oxidation states and band gap positions which was 2.67 eV for the composite and gave the red shift in spectra. The photocatalytic activity of FeS_2 and TiO_2 was tested alone and also for composite but the photoactivity of nanocrystal composite for the degradation of methylene blue, was much higher than that of pristine compounds. Experiments were performed by varying the parameters like pH, catalyst dose and pollutant concentration. Composite was found to be much stable as it shown only 9% loss in photocatalytic activity after 3 cycles of use [142].

Among venedates, bismuth vanadate and indium vanadate have gained much respect mainly due to stability, nontoxicity and photocatalytic activity under visible light (λ > 420 nm) in coupled or single conditions. For coupled semiconductor heterostructure of TiO_2/$InVO_4$, $InVO_4$ was prepared using organic precursor method and was incorporated with TiO_2 by grinding, the catalyst shown higher photocatalytic degradation of 2-CP under visible light as compared to TiO_2 alone. Among different conditions the best condition for degradation were found at 5pH in 1 g/L of catalyst and 50 mg/L of pollutant with irradiation time of 180 min [143, 144]. Li et al. has reported the unique structure of TiO_2 incorporated on the polymethyl methacrylate (PMMA) nanofibers for the degradation of pollutants under the UV irradiation. The TiO_2 nanoparticles were supported on fibers by the hydrothermal treatment at 135°C for 8 h using titanium n-butoxide and PMMA nanofibers as precursors. Analysis indicated that about 42% of tetragonal anatase TiO_2 particles were adsorbed on the surface of fibers (**Figure 21**). Methylene orange was used as model pollutant and its complete degradation was observed after 50 min using 0.1 g of catalyst for 10 mg/L of pollutant. This good photocatalytic activity could be attributed to the (i) decoration of TiO_2 nanoparticles on the PMMA fibers which assisted the adsorption of pollutants and helped them to come in contact with photocatalyst; (ii) high quantity of adsorbed TiO_2 provides numerous action sites for the degradation of pollutants. Catalyst revealed the efficiency of about 94.4% even at the end of fifth cycle for the degradation of fresh MO which determined the stability and reusability of catalyst [145].

A novel nanocomposite of BiOBr/TiO_2 has been reported and prepared for the first time with facile acid assisted precipitation method by using TTIP, KBr and $((BiNO_3)_3.5H_2O)$ as precursors. Catalysts of different percentages were prepared by using different molar ratios of TiO_2 as a base material and 15% BiOBr/TiO_2 behaved as an active catalyst. Analysis and characterization revealed the perfect structural integrity, increased surface area and increased absorption of light because of the formation of nanoscale layered butterfly like clustal structures with each flake having the broad absorption spectrum of about 500–800 nm with <50 nm thickness (**Figure 22**). PL spectral analysis determined the band gap of 2.81 eV and wavelength was observed from 480 to 570 nm for 15% BiOBr/TiO_2 and it gave the

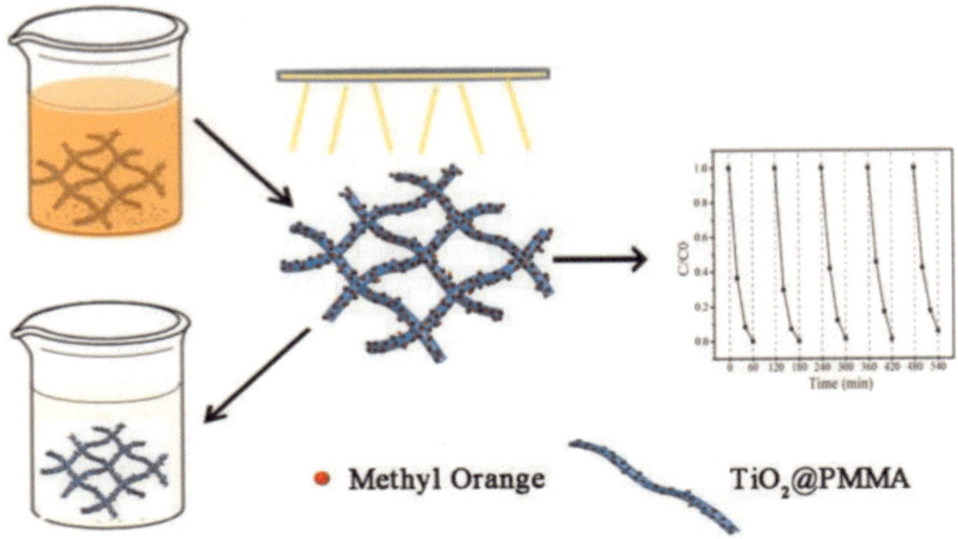

Figure 21.
Schematic illustration of formation of PMMA fibers and their photoactivity [145].

degradation of aqueous ciprofloxacin (target pollutant) under visible light at $\lambda > 420$ nm about 92.5% and under direct sunlight it gave 100% degradation utilizing reaction time of 2.5 h. Radical scavenging study indicated that besides superoxide radicals and holes the $OH^{\bullet}$ had dominant effect on net photodegradation process. Results also suggested that the catalyst was 9.4 and 5.2 times active in degradation than pristine BiOBr and TiO_2, reusability experiments determined the high stability of catalyst as it suffered loss of only 12% even after 5 cycles of fresh CIP degradation [146].

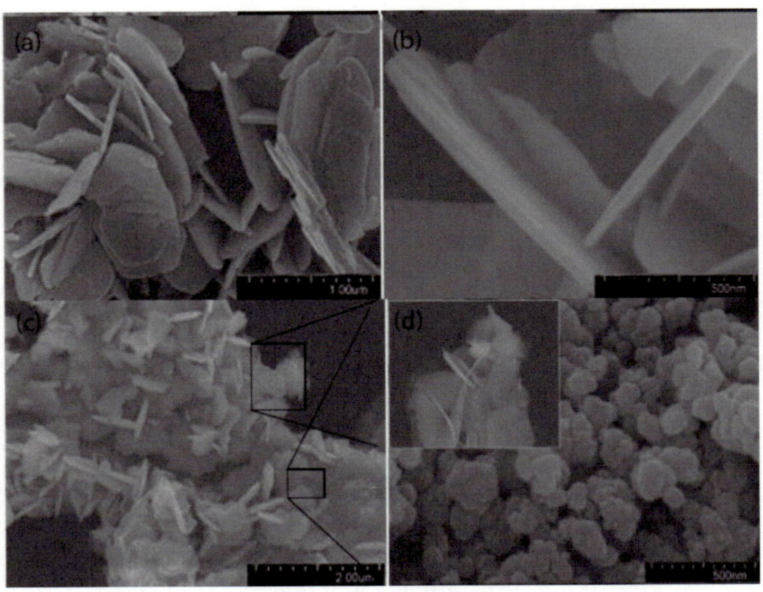

Figure 22.
SEM images of (a, b) pristine BiOBr; and (c, d) 15%BiOBr/TiO_2 nanocomposite.

4.2 Heterostructures/nanocomposites of different SCs

In addition to TiO_2 there are numerous semiconductors which can be coupled with other SC and can be incorporated into heterostructures or composites like WO_3, ZnO, $BiVO_4$, ZnSe, C_3N_4, In_2O_3, CuS, MoS, Cu_2O and many more. So another example of inorganic heterostructure is the template free simple synthesis of CdS-ZnO nanocomposites. This self-assembled flower like structure resulted from coupling of two SCs which increased the charge separation and demonstrated greater photocatalytic activity. The average size of flower was 400 nm with the petal size of 100–150 nm and the rod among petals had the size of 10 nm, band gap of composite was analyzed to be 2.19 eV and absorption wavelength of 561 nm was reported. The composite gave the photodegradation of rhB about 90% in 190 min under solar light and 24% in dark. The scheme of process of particles is shown in (**Figure 23**) [147].

Cho et al. prepared three different types of 3D ZnSe/ZnO heterostructures by simple solution-based surface modification reactions. Three different types of heterostructures demonstrated higher photocatalytic activities by exhibiting absorption in visible region (at $\lambda > 486$ nm) as compared to pure 3D ZnSe and ZnO structures, the visible light activity of heterostructures varied according to the crystal structures. As shown in (**Figure 24**) the CB of ZnO is in between the CB and VB of ZnSe showing the type II band alignment, before exposure to sunlight the catalysts and orange II (targeted pollutant) were kept at stirring for 30 min in dark to attain an equilibrium for adsorption and desorption, upon exposure to visible light the e^-/h^+ pair was generated in ZnSe crystal, electron transferred to the CB of ZnO to reduce oxygen and holes could oxidize either the water or OH ion or directly oxidize the orange II, ended with the mineralization of organic compound [148].

Zeyan et al. has produced the In_2O_3/ZnO heteronanostructures by the co-precipitation method having average size of 40–60 nm by using the respective precursors. The composite was annealed at different temperatures ranging from 600 to 1000°C, among changed compositions and annealing temperatures the highest photocatalytic activity for the degradation of methylene blue was shown by the composite annealed at 800°C with In/Zn molar ratio of 1:1 and the maximum absorption was about $\lambda = 663$ nm. A p-n junction was established at the SCs interface and presence of Zn^{2+} and In^{3+} was reported in lattice, that enhanced the charge separation and production of $OH^{\bullet}$ and $O_2^{\bullet-}$ attributed as the main reason for enhanced photocatalytic activity semiconductor coupling ([149]). Coupling of

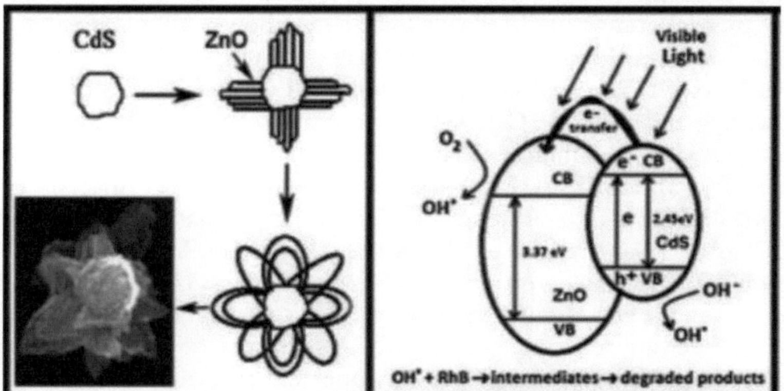

Figure 23.
Structure of flower shaped nanocomposite and admirable charge transfer due to heterostructure formation [147].

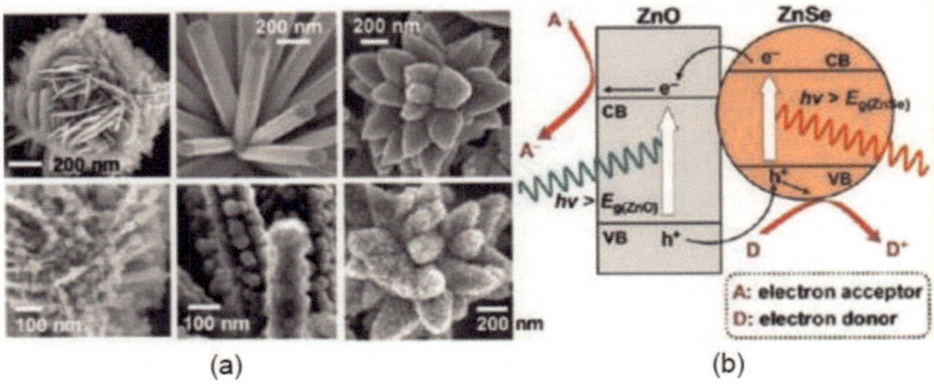

Figure 24.
(a) SEM images of different 3D ZnO/ZnSe heterostructures and; (b) schematic diagram of charge separation by coupling of two SCs [148].

mixed oxide SC with metal oxide has also been reported and have shown some modified properties, example for this type of heterojunction is the nanocomposite of ZnO-$ZnWO_4$ (both are the n-type SCs) prepared by the sol-gel method and was analysed by XRD, BET, SEM and TEM. The nanocomposites were prepared by varying the molar ratio of ZnO and $ZnWO_4$ to study the influence of molarity on photoactivity and Zn-$ZW_{0.25}$ calcined for 2 h at 600°C shown greater photodegradation of 4-nitrophenol (target pollutant). The band gaps for ZnO and $ZnWO_4$ was calculated to be 3.21 and 3.14 eV, when the nanocomposite was exposed to UV light the electron transferred from the CB of ZnO (E_{CB} = −0.36 eV) to CB of $ZnWO_4$ (E_{CB} = −0.14 eV) due to difference in the position of CBs (from cathodic to anodic condition) and same behaviour was proposed for hole to transfer from VB of $ZnWO_4$ (E_{VB} = 3.00 eV) to VB of ZnO (E_{VB} = 2.84 eV), thus enhancing the charge separation and completion of circuit due to formation of heterostructures which resulted from the mixing of two SCs with dissimilar energy levels of VB and CB [150]. In addition to the above mentioned heterostructures other structures like nanorods, nano sheets, nanowires, CdSe/ZnO, CdsSe/ZnO [17], TiO_2/ZnO [151], CuS/ZnO [152] has also been found to enhance visible light activity either for pollutant degradation, overall water splitting or hydrogen production.

BiOCl/$BiVO_4$, another novel p-n heterostructure of coupled semiconductors with type II band alignment was prepared by hydrothermal method for the decomposition of methyl orange under visible light irradiation having ability for active separation of charges (**Figure 25**), the powders were prepared by using NH_4VO_3, Bi $(NO_3)_3{\cdot}5H_2O$, NaOH and HCl as precursors for $BiVO_4$ and BiOCl, followed by the stirring, heating and washing. Catalysts having different molar concentrations were prepared by varying the amount of HCl in the mixture and the catalyst having the molar ratio of 0.75BiOCl/$BiVO_4$ was seemed to be more active. Analysis and characterizations determined th-presence of Bi as Bi^{3+} and vanadium as V^{5+} and O as $O_2^{\bullet-}$ and Cl as Cl^-. For the tests of photocatalytic activity, Degussa P25 was used as control group, and results determined the increased degradation of MO in BiOCl/$BiVO_4$ system and it was about 1.89 times greater than that of TiO_2 and 3.54 times from $BiVO_4$. Free radical scavengers illuminated that the holes in the valence band of *p*-type part of heterojunction played an important role in the direct oxidation of MO, $OH^\bullet$ were the dominant species and DO acted as electron acceptor from the CB. The stability of catalyst was tested and after the use of 5 cycles, it did not show any significant loss, which indicated the effectiveness and stability against photocorrosion [153]. Zhijie et al. has reported the fabrication of a novel

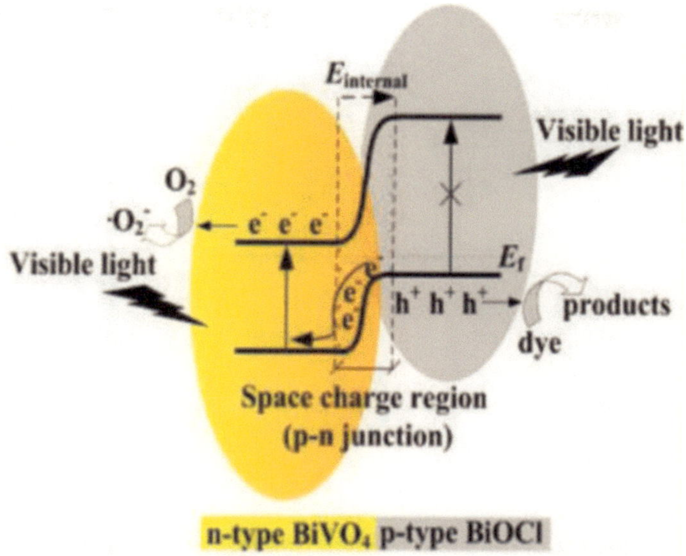

Figure 25.
Schematic illustration of effective charge separation between n- and p-type SC upon exposure to UV and Vis light [153].

nanocomposite of Bi_2S_3/Bi_2WO_6 by hydrothermal method to reduce the limitation of low photocatalytic ability of Bi_2WO_6 resulted from the potential relaxation of electron into hole. In this structure Bi_2S_3 acted as a sensitizer having the band gap of 1.3 eV and the system possessed the photoabsorption of 800 nm which meant it covered almost whole range of visible region. The equipped photocatalyst exhibited much enhanced photocatalytic activity for the degradation of phenol (a colorless pollutant) under visible light which was about 6.2 times greater than that of original Bi_2WO_6. Thanks to efficient charge separation for the enhanced photocatalytic activity resulted from the formation of heterojunction at the semiconductor interface [82, 83]. The heterostructure formed by mixing of graphene with $BiFeO_3$ was prepared by one-pot hydrothermal method. The band gap of composite could be fabricated from 1.78 to 2.24 eV by changing the concentration of OH groups during the synthesis. Raman and XPS analysis revealed that the formation of Fe-O-C bonds enhanced the stability of composite. The increased photodegradation of Congo Red (CR) dye was observed in 2 h which increased from 40 to 71% due to the breakage of azo bonds and naphthalene rings at different wavelengths which can be due to the increased adsorption of CR on the graphene due to the infinite numbers of conjugated π-bond ([122]).

WO_3 is a known semiconductor due to its polycrystalline forms and smaller band gap of about 2.4–2.8 eV depending on crystal structure; which means its electrons can be excited by visible light irradiation, the main limitations of this SC which keep it far from becoming a useful semiconductor are; higher rate of recombination and its unstability against photocorrosion [154]. WO_3 is incorporated in heterojunctions in order to increase the absorption in visible light region. Shamaila et al. has reported the formation of WO_3/BiOCl heterojunction or the degradation of organic pollutants under visible light using $BiCl_3$, Ammonium tungstate and ethanol as precursors, 2D nanoflakes of BiOCl of the size of 75–200 nm were fabricated (**Figure 26**) by following a new low temperature route and nanocomposite was prepared to enhance the visible light response, BiOCl performed the role of a main photocatalyst while WO_3 worked as a sensitizer. For the prepared composite the absorption range increased from 360 to 500 nm as the quantity of WO_3 increased

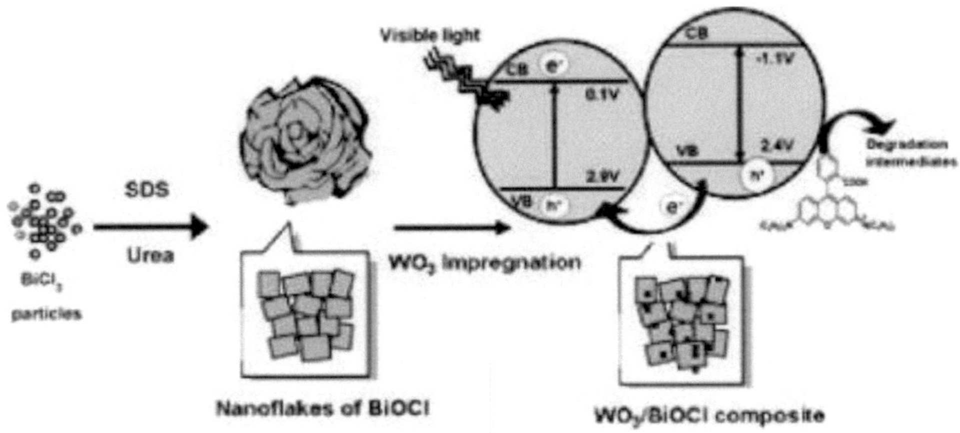

Figure 26.
Composite synthesis route and proposed mechanism for photodegradation of rhB [155].

while synthesis, the catalyst gave 100% deterioration of Rhodamine B was under visible light during the irradiation time of 180 min, which was greater than the activity of WO_3, BiOCl and Degussa P25 [155].

Another heterojunction photocatalyst with high photocatalytic activity is $WO_3/SrNb_2O_6$, was prepared by the milling-annealing method which was found to be better than the direct mixing method as this method could build a firm chemically bonded interface between two semiconductors. Results determined that the composite had higher photocatalytic activity as compared to pure WO_3 and $SrNb_2O_6$. As we know that anything which can scavenge the excited electron can be a reason for increased photocatalytic process so, the effective charge separation among semiconductors and formation of holes and radicals led to the direct or indirect degradation of methyl orange [156]. Ag_3PO_4/WO_3 nanocomposites were prepared by a deposition-precipitation method using $Na_2WO_4{\cdot}2H_2O$, NH_4Cl, $AgNO_3$ and Na_2HPO_4 as precursors and prepared powders were characterized by SEM, XRD and UV-Vis. Catalysts were prepared of variable molar ratios of Ag_3PO_4:WO_3 and the catalyst having ratio of 6:4 ($AW_{6/4}$) shown the greater photocatalytic activity for the degradation of rhB and MO at the wavelength $\lambda > 420$ nm, mainly due to the excellent separation of charge carriers. Reusability tests indicated that composite had higher recyclability as compared to Ag_3PO_4 alone as $AW_{6/4}$ gave 97% removal of rhB and MO in 6 and 35 min (after being exposed to visible light) even after the fifth run (had same efficiency for first run) while Ag_3PO_4 gave only 25% degradation at the fifth run due to the absence of sacrificial donor. Stability analysis revealed that catalyst $AW_{6/4}$ was stable against photocorrosion as it retained the XRD pattern even after 5 cycles for degrading fresh rhB, only a small amount of metallic Silver was observed on Ag_3PO_4 after use, SEM analysis also supported stability result [157].

A hybrid photocatalyst $W_{18}O_{49}/TiO_2$ having an urchin like structure (**Figure 27**) was prepared by an alcohol thermal method, which had high surface area of 178 $m^2\ g^{-1}$ and shown absorption in the wide range of 200–800 nm and tested for the photodisintegration of MO and phenol under UV-visible irradiation. The hybrid photocatalyst attained synergetic increase in photoactivity and photostability of $W_{18}O_{49}$, well-related band structure enhanced the charge separation and transfer which led to the production of free radicals as well as prevented the $W_{18}O_{49}$ from self-oxidation due to holes as they moved towards TiO_2. Free radical scavenging tests indicated that $O_2^{\bullet-}$ was reactive oxidizing specie for the degradation of MO and phenol. Reusability tests revealed the stability of catalyst as it maintained its

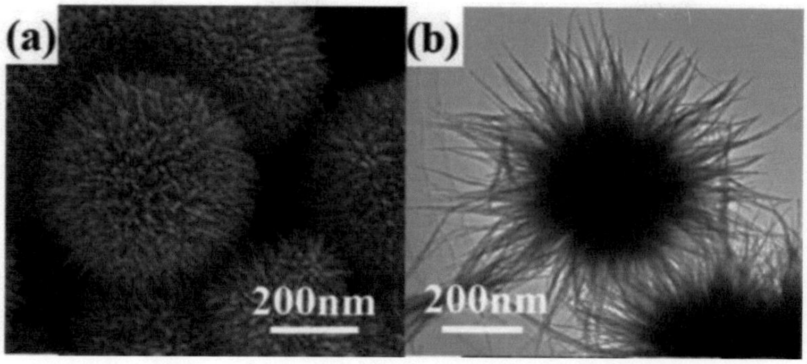

Figure 27.
(a) SEM image of $W_{18}O_{49}$ (b) TEM image of $W_{18}O_{49}$ [158].

XRD spectra even after the fifth cycle which meant catalyst did not suffer any noticeable loss, hybrid composite also maintained its original blue color at the end of fifth run [158].

From metal tungstate an example of a p-n heterojunction is the fabrication of a novel Z-scheme $WO_3/CdWO_4$ photocatalyst by the fusion of sheet like WO_3 (n-type) and rod like nanostructure of $CdWO_4$ (p-type), for this purpose the catalyst were prepared separately by using respective precursors and fusion was done by using hydrothermal and chemisorption methods. As-prepared catalysts were used for the degradation of organic dyes like Methly orange, Rhodamine B and methylene blue, upon exposure of light the transfer of electrons and holes occurs from the $CdWO_4$ to WO_3 and vice versa which led to circuit completion and efficient increase in charge separation, PL spectra gave the smaller curve peak proving the less relaxation of electron towards VB and the absorption intensity of $\lambda > 476$ nm. Stability tests were performed to check the structural reliability of catalyst, it did not show any significant loss even after 3 cycles of reuse for the degradation of fresh dyes. Results also indicated the degradation of dye was about 7 and 2.3 times greater than $CdWO_4$ and WO_3 alone which was credited to enhanced surface area and effective separation of charge carriers [159].

In addition to the binary composites, some tertiary (complex) composites have been reported, where the multi-photon excitation took place in the photoactive materials and charge transfer and separation was increased due to increased surface area and different transition states. $KTaO_3$-CdS-MoS_2 is an example of tertiary composite which was prepared by hydrothermal method by rigorously following the temperature and pressure conditions, the obtained powders were of different structures like nanoleaf, cubic and hexagonal spheres exhibiting activity under both visible and UV light and were used for the degradation of toluene and phenol. The catalysts shown 42% degradation under visible light and 80% degradation of pollutants under UV light. Prepared catalyst exhibited good stability (about 50%) even after 4 cycles for the degradation of fresh toluene [160]. In case of some ternary composites like ZnO-AgBr-Ag_2CrO_4 (n-n junctions), two SCs worked as donors of photoexcited electrons and oxidation of pollutants occurred by the formation of superoxide radicals at the CB. By the charge separation the phocatalytic activity of composite was 16 and 7 times higher than those of ZnO and ZnO/AgBr for the degradation of rhB [161]. Rhodamine B also suffered degradation by the photocatalytic activity of In_2O_3-AgBr-$BiWO_6$ (**Figure 28**), prepared by microwave assisted irradiation method, optical, morphological and structural properties were analysed by XRD, SEM, TEM, XPS, HRTEM which indicated the flower-like pattern of assembled nanoparticles. The increased photoactivity under UV, visible and solar

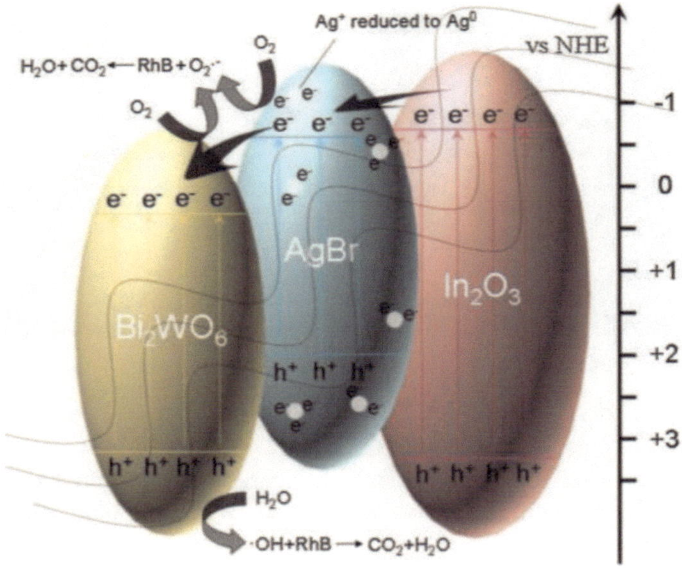

Figure 28.
Proposed mechanism for the degradation of rhB on ternary composite [162].

light was attributed to increased surface area and charge separation due to differences in the redox potentials of SCs bands. Reactive species scavenging testes determined that OH$^{\bullet}$ and superoxide radicals were dominant reactive species [162].

Besides all of the above discussed heterostructures, nanocomposites, coupled semiconductors, either oxides or sulfides, there are numerous other heterojunctions like $WO_3/NiWO_4$ [163], Ag_3PO_4/Bi_2MoO_6 [164], $Ag_3PO_4/AgBr$ [165], CdS/Bi_2MoO_6 [166], CdS/Ta_2O_5 [167], BiOBr/$ZnFe_2O_4$ [168], $Cu_2O/SrTiO_3$ ([169]), ZnS/CuS/CdS [170], $SnO_2/ZnO/ZnWO_4$ [171] and many more have been reported to increase the ability of pure SCs to harvest the sun light either in the field of water or wastewater treatment for the overall splitting of water, degradation of Organic, inorganic pollutants, phenols, MCs, POPs and other emerging contaminants.

5. A glance at g-C_3N_4-based heterostructures/nanocomposites

It has been more than a decade that people are being aware from environmental protection and conservation, the era of industrialization besides increasing the gross productivity also posed serious effects on the health of water bodies by discharging unchecked amounts of effluents and other chemicals, in order to improve the water quality the focuses are being paid to treat water and wastewater, the idea of semiconductor photocatalysis has been emerged which uses solar energy in its full potential, for this purpose many SCs either in single or in heterostructures have been discussed in Section 6.

In addition to first generation photocatalysts, an emerging polymeric photocatalyst is Graphitic Carbon Nitride (g-C_3N_4) which is an earth copious visible light active catalyst, having unique 2D structure with high stability and flexible structure (which can be tailored) and low band gap of 2.7 eV. Pure gCN have high rate of recombination, smaller surface area and low ability to exploit visible light thus reducing the quantum efficiency of photosystem, the VL activity of gCN has been reported to increase by making heterostructures with other SCs by taking the advantage of its layered structure which assisted in hybridization with other constituents like CdS, TiO_2, $ZnWO_4$, ZnO, etc. either for the degradation of pollutants or for the evolution of O_2 or H_2 gas [172]. Some examples of gCN HSs/HJs/Composites are discussed below.

Novel $g\text{-}C_3N_4/TiO_2$ composites were prepared by the facile sonication method using melamine and titanium tetrachloride as precursors followed by stirring, sonication and drying at room temperature. Different catalysts were prepared by varying the concentrations of precursors and $g\text{-}C_3N_4/TiO_2$-1.5 was found to be more active among other catalysts, as it shown higher photocatalytic degradation of MB under UV and visible light. Results demonstrated that catalyst shown 6.92 and 2.65 time greater activity than pure gCN and TiO_2 under UV light, while 9.27 (gCN) and 7.03 (TiO_2) folds greater photoactivity under visible light, which can be attributed the increased visible light absorption and efficient charge transfer in composite at interfaces of SCs. In the hybrid catalyst gCN could be triggered under visible light and the photoexcited electrons of gCN transferred to the CB of TiO_2 due to interfacial connections and differences of redox potentials [173]. Tunable band gap of some semiconductors is an effective way to harness solar light and among those gCN is a known SC. Two SCs having smaller band gaps (gCN and Bi_2MoO_6) were coupled to prepare a Z-scheme nanocomposite, number of catalysts were prepared by varying the concentrations of precursors and 25% $g\text{-}C_3N_4/Bi_2MoO_6$ was seemed to be most active, structure, morphology, light absorption spectra and charge carriers separation efficiency was analyzed and photocatalytic activities were evaluated for the degradation of MB. Both SCs in composite were excited at the λ = 410 nm, due to the negative redox potentials of gCN (as compared to Bi_2MoO_6), excited electrons of gCN transferred to CB of Bi_2MoO_6 and holes of later should had been migrated to the VB of gCN, but by this route the holes at the VB of gCN could not react with water of OH ion near its surface and same for Bi_2MoO_6, so it was observed that the excited electrons transferred to their respective CB, the relectrons from the CB of Bi_2MoO_6 were transferred to the VB of gCN (spatial separation) which inhibited the local recombination (**Figure 29**) thus completing the Z-scheme route, the CB of Bi_2MoO_6 and VB of gCN could not react with molecules in their vicinity and the catalyst shown 4.8 and 8.2 folds greater photoactivity than pristine SCs [174].

Feng et al. reported the fabrication of S-doped $g\text{-}C_3N_4/Au/CdS$ composite as Z-scheme photocatalyst where CdS was deposited on g-CNS system in which Au nanoparticles were sandwiched between g-CNS and CdS (two visible light responsive SCs) by the chemical bath deposition and worked as charge transporter (**Figure 30**), whole system increased the degradation of rhB, MO, MB and increased

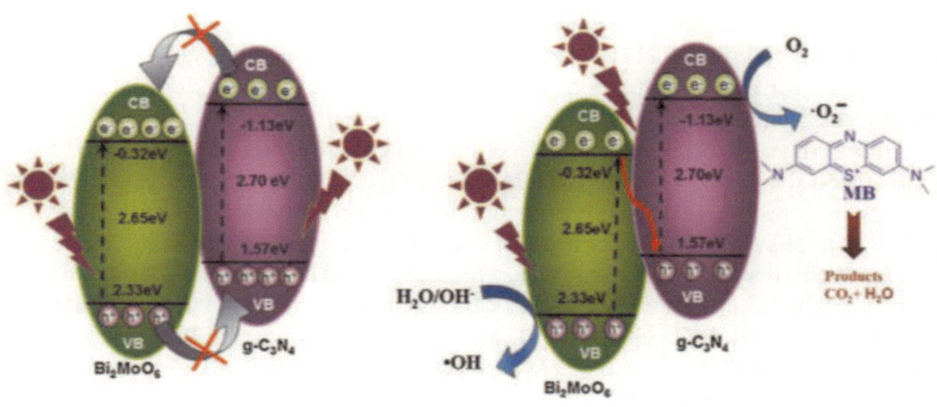

Figure 29.
Proposed mechanism of charge transfer in Z-scheme photocatalyst [174].

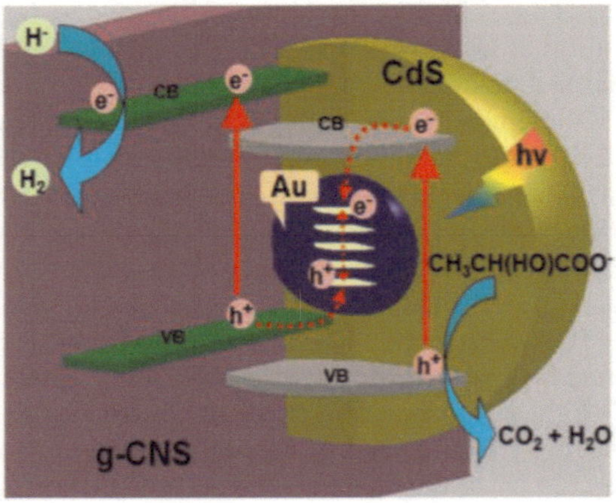

Figure 30.
Proposed mechanism of Z-scheme charge separation and transportation in g-CNS/Au/CdS [175].

reduction of water to hydrogen was observed under visible light (λ = 560 nm). The Z-scheme photocatalyst was characterized by SEM, HRTEM, PL spectra, Vis-diffuse spectra. These elementary, crystal and microcrystal analysis demonstrated the uniformity of novel composition, results revealed that the catalyst exhibited great photocatalytic activity as compared to g-CNS, g-CNS/Au, g-CNS/CdS, and pure CdS [175]. Novel organic-inorganic composite of g-C_3N_4-CdS has been prepared which gave the degradation of organic pollutants under visible light. Results indicated the higher photocatalytic activity of about 20.5 and 3.1 times for degradation of MO as compared to gCN and CdS alone and 41.6 and 2.7 times higher for the degradation of 4-aminobenzoic acid [176].

GCN has also been reported to be composed with 2D graphene which is being focused due to its unique properties like amazing thermal, mechanical, surficial and electric properties. Tong et al. has prepared 3D porous gCN/graphene oxide aerogel structure by using gCN sheets and graphene oxide as precursors by following hydrothermal treatment method this composition resulted to enhanced visible light absorption, decreased charge relaxation and increased adsorption capacity. Analysis and characterizations were performed to check intrinsic properties. As-prepared catalyst was used for the CO_2 evolution and MO degradation which was about 92% in 4 h under visible light [177]. Han et al. has reported different g-C_3N_4/graphene nanocomposites which were proved to be effective for the degradation of dyes under visible light, different hybrids of gCN/graphene were prepared by using different methodologies and synthesis processes like solvothermal treatment, hydrothermal treatment, electrochemical reactions, etc. and their photoactivity was investigated for the hydrogen, oxygen, CO_2 evolution and degradation of different organic compounds [178].

Among ternary composites the novel magnetically recoverable composite of g-C_3N_4/Fe_3O_4/$NiWO_4$ was prepared by refluxing calcination method, different hybrids were prepared by varying the precursor's concentration among which gCN/Fe_3O_4/$NiWO_4$ (30%) was found to be more active, catalysts were characterized for their morphology, texture, electronic, thermal and magnetic properties. Hybrid catalyst shown their high photocatalytic activity under visible light for the degradation of rhB, MO, MB, fuchsine and phenolic compounds. Results confirmed that formation of HJs between gCN/M/$NiWO_4$ increased the surface area and charge separation (**Figure 31**), thus increasing the quantum efficiency of system, photocatalytic activity of gCN/M/$NiWO_4$ (30%) was 12, 30, 52 and 6 folds greater for rhB, MB, MO, fuchsine and phenol as compared to gCN, Fe_2O_3 and $NiWO_4$

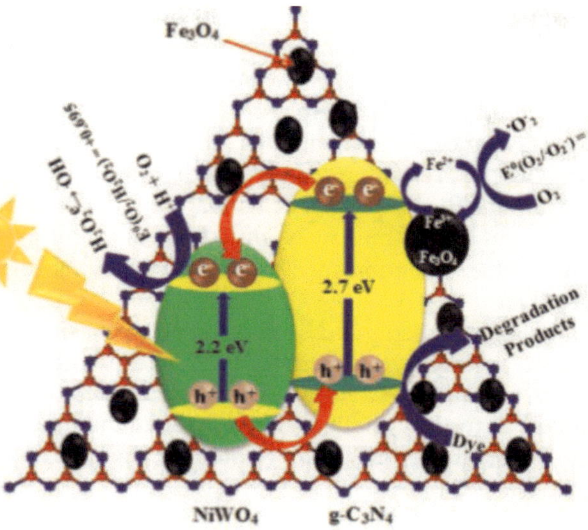

Figure 31.
Proposed mechanism of charge transfer in gCN/M/NiWO$_4$ nanocomposites [179].

alone. Additionally the prepared composites exhibited high degree of stability even after the 4 cycles of reusability as it degraded about 87% of rhB at fourth cycle, unluckily, nanohybrid suffered from reduced surface area after the last run which was attributed to the blockage of degradation reaction intermediates at surface sites [179]. Zhang et al. has used one-step solvothermal method for the preparation of spinel $ZnFe_2O_4$ nanoparticles and its decoration on g-C_3N_4 sheets for formation of water soluble magnetic-functionalized g-C_3N_4/$ZnFe_2O_4$. The magnetic properties of composite were controlled by the size of composite and decorated quantity of $ZnFe_2O_4$. Results indicated that catalyst exhibited efficient photocatalytic activity for degradation of MO under visible light and it was mainly due to the synergist effect of both SCs like smaller particle size and high solubility in water, interestingly the catalyst could be separated from aqueous solution by magnet. CN-ZnFe showed 98% of degradation of MO in 180 min which was 6.4 and 5.6 times higher than g-C_3N_4 and $ZnFe_2O_4$ alone. Free radical scavenging test revealed the decrease in efficiency indicated the vital role of $OH^•$ and superoxide radical in the pollutant degradation. The increased photocatalytic activity was credited to sufficient charge separation ([180]). Synthesis of novel magnetic CdS/$ZnFe_2O_4$ nanocomposites has been reported (by two-step hydrothermal method) in order to enhance photocatalytic activity and photostability of SCs under visible region [181].

Heterojunctions of g-C_3N_4 with carbonaceous particles has been reported to make recyclable, stable and more efficient photocatalyst for the water treatment purposes. Chai et al. synthesized the g-C_3N_4 modified with fullerene for the removal of dyes from water under visible light. Fullerenes (C_{60}) are known to have 30 molecular orbits with 60 π electrons with closed shell structure which helped in separation and transport of charge carriers, especially the photoexcited electrons. Because of the larger surface area and unique morphology, the electrons excited from g-C_3N_4 shifted on C_{60} thus got separated from hole and lessened the electron relaxation towards hole, the electron hole pair separation resulted in the opening of benzene ring during the process of organic molecule degradation [182, 183]. Bai et al. prepared g-C_3N_4/fullerene composite using thermal polymerization of dicyandiamide in the availability of C_{60} at 550°C. Addition of C_{60} in the gCN, shifted the VB of gCN at lower energy level thus its oxidation potential under visible light was increased. **Figure 32** demonstrate the increased photocatalytic activity due to effective electron hole pair separation, when the catalyst was exposed to visible light electrons got excited from VB (formed by N2p orbitals) to CB (formed by 2Cp orbitals) of g-C_3N_4 leaving the holes behind in the VB, the excited electrons jumped to C_{60} from CB of gCN thus retaining the separation. Results demonstrated the enhanced degradation of MB and phenol at $\lambda > 420$ nm, degradation was about 2.9 and 3.2 time greater than that of bulk gCN, due to the efficient production of $OH^{\bullet}$ from holes, which acted as oxidation species and played a vital role in opening of benzene rings ([184]).

In addition to above discussed g-C_3N_4-based nanocomposites, there is a lot of literature based on the synthesis, uses and types of heterostructures based on the multifunctional approaches, as g-C_3N_4-based nanostructures (either with oxides or sulfides, either binary or tertiary) have become the main member of semiconductor photocatalysis family for the degradation of pollutants in environment specially in water. Besides g-C_3N_4/Fe_2O_3, g-C_3N_4/CeO_2 g-C_3N_4/MoO_3, g-C_3N_4/Fe_3O_4, g-C_3N_4/$Ni(OH)_2$, g-C_3N_4/Ag_2O, g-C_3N_4/MoS_2, g-C_3N_4/NiS, g-C_3N_4/TaON, g-C_3N_4/ZnO, g-C_3N_4/g-C_3N_4, g-C_3N_4/In_2O_3, g-C_3N_4/WO_3, SiO_2/g-C_3N_4 and some ternary composites like, gCN/Fe_3O_4/$CuWO_4$, BiOCl/Bi_2MoO_6/g-C_3N_4, Ag_2CrO_4/g-C_3N_4/GO, g-C_3N_4/TiO_2/CNT, G-C_3N_4/CeO_2/ZnO there are innumerable gCN composites, some composites with their photocatalytic activity and efficiency under visible light are mentioned in **Table 2**.

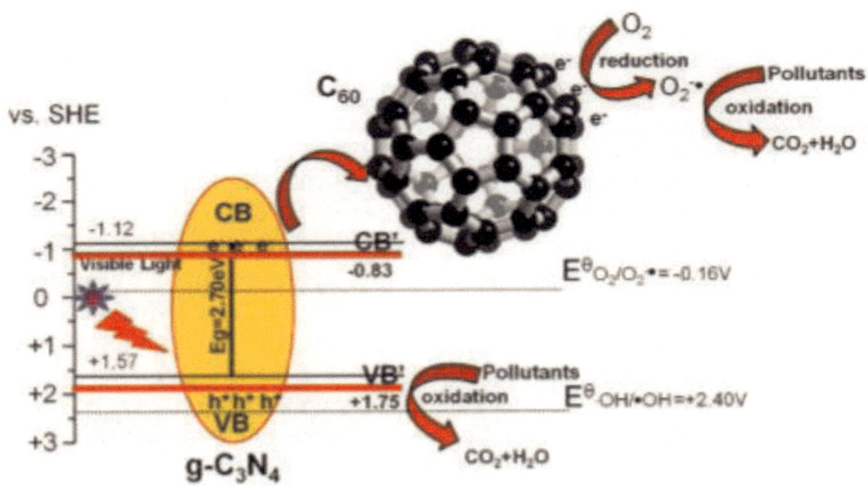

Figure 32.
Schematic illustration of charge separation and decomposition mechanism of organic pollutants by production of free radicals ([184]).

Composite	Precursor/synthesis method	Targeted pollutant	Degradation efficiency/reason under visible light	References
TiO_2/g-C_3N_4	P-25 and dicyandiamide	Methylene blue	90% degradation in 300 min, 3.5 greater efficiency than SC alone.	[185]
WO_3/ g-C_3N_4	Dicyandiamide and $(NH_4)_5H_5[H_2(WO_4)_6]\cdot H_2O$	MB and 4-CP	Degraded 97% of MB and 43% of 4-CP	[186]
$CdWO_4$/ g-C_3N_4	Melamine/mixed calcination synthesis	rhB	Degradation of rhB due to $O_2^{\bullet-}$ radical formation, 1.6 and 54.6	[187]

Composite	Precursor/synthesis method	Targeted pollutant	Degradation efficiency/reason under visible light	References
			times greater efficiency than gCN and CWO.	
In_2S_3/ g-C_3N_4	Urea/hydrothermal synthesis	rhB	HJ formation widened the light absorption spectra, increased EHP separation and $O_2^{\bullet-}$ were reactive OS.	[188]
V_2O_5/ g-C_3N_4	Melamine/one-pot synthesis	rhB	Deterioration of rhB due to $O_2^{\bullet-}$ and $OH^{\bullet}$, charge separation increased quantum efficiency under visible light.	[189]
ZnO/ g-C_3N_4	Melamine and zinc acetate/simple calcination method	MO and p-nitrophenol	Degraded 97% of MO in 80 min, which was three and six times greater than gCN and ZnO alone.	[190]
$ZnWO_4$/g-C_3N_4	Melamine/facile chemisorption	MB	Higher electron hole pair separation, OH and $O_2^{\bullet-}$ were reactive species	[191]
BiOI/ BiOBr/g-C_3N_4	Dicyandiamide	MB	Effective charge separation, production of $OH^{\bullet}$, $O_2^{\bullet-}$ and h^+	[192]
ZnO/ZnS/g-C_3N_4	Melamine/two-step facile synthesis	H_2 Production	Hydrogen production 1205 μmol/g-C_3N_4:16 μmol/g	[193]
YVO_4/g-C_3N_4/Ag	Hydrothermal/urea deposition method	MO	3 times higher activity as compared to gCN alone	[194]
C_3N_4/CNT/ NiS	Sol-gel method/direct precipitation of cyanamide	H_2 Production	148 times higher activity than pure gCN/CNT	[195]
Ag_3PO_4/ GO/g-C_3N_4	Dicyandiamide/ precipitation	RhB degradation	94.8% degradation in 50 min.	[196]
g-C_3N_4/AgBr/ RGO	Urea/hydrothermal treatment	Tetracycline and 2,4-DP	78.4% degradation in 90 min and 68.2% within 6 h.	[197]
g-C_3N_4/TiO_2/ ZnO	Hydrothermal treatment	p-toluene sulfonic acid	90% degradation by composite, while only 40% by gCN	[198]
g-C_3N_4/ CeO_2/ZnO	Urea/pyrolysis and exfoliation method	MB	11 times greater degradation than bare gCN	[199]
Ag/g-C_3N_4/ $NaTaO_3$	Melamine/photo deposition process	Dyes and tetracycline	Improved dye degradation and 95.47% removal of tetracycline.	[200]

Table 2.
g-C_3N_4-based heterostructures/nanocomposite.

6. Conclusions

Majority organic, inorganic and other numerous pollutants which are either in the records of EPAs or not, are the refractory compounds and tend to become the part of fresh or potable water thus making it unfit for use. Besides employing membrane systems, VLA semiconductor photocatalysis has gained fame due to their cost effectiveness and eco-friendly nature and becoming a promising phenomenon for the treatment of water on global scale. The rapid development in field of nanotechnology has opened the doors for various advanced photocatalytic systems which includes the formation of complex structures by doping, co-doping, heterostructures, 0D to 3D structures and nanocomposites as heterogeneous photocatalysts, has been developed in recent decades and has helped the environment and water to get rid of stubborn pollutant species, such as bacteria like *E. coli*, *B. subtilis*, giardia cysts, harmful cyanobacteria and their toxins like microcystins LR, RR, YR, LA (types of Microcystins), etc. which are harmful for different life forms. However, the application of visible light active semiconductors in the long run will be a reason of development of sustainable environment, due to development of advanced oxidation processes which are triggered by sunlight, will become a renewable energy source for example the evolution of hydrogen, oxygen and carbon dioxide gas. Therefore, an effective assessment of SCs nanomaterials is still required to test true photoactivity and to enhance the commercialization for future regards.

Author details

Fatima Imtiaz[1], Jamshaid Rashid[1,2]* and Ming Xu[2]

1 Department of Environmental Science, Faculty of Biological Sciences, Quaid-i-Azam University, Islamabad, Pakistan

2 Key Laboratory of Geospatial Technology for the Middle and Lower Yellow River Regions, College of Environment and Planning, Henan University, Jinming Avenue, Kaifeng, China

*Address all correspondence to: jamshaidrashid@gmail.com; jrashid@qau.edu.pk

References

[1] Serpone N. Advanced oxidation processes. In: Kirk-Othmer Encyclopedia of Chemical Technology. Vol. 1. Toronto, Canada: John Wiley and Sons, Inc.; 2000. pp. 1-17

[2] Fujishima A, Zhang X, Tryk DA. TiO_2 photocatalysis and related surface phenomena. Surface Science Reports. 2008;**63**:515-582. DOI: 10.1016/j.surfrep.2008.10.001

[3] Mills A, Le Hunte S. An overview of semiconductor photocatalysis. Journal of Photochemistry and Photobiology A: Chemistry. 2000;**108**:1-35

[4] Ohtani B. Photocatalysis A to Z—What we know and what we do not know in a scientific sense. Journal of Photochemistry and Photobiology, C: Photochemistry Reviews. 2011;**11**: 157-178. DOI: 10.1016/j.jphotochemrev.2011.02.001

[5] Goodeve CF, Kitchener JA. Photosensitisation by titanium dioxide. Transactions of the Faraday Society. 1938;**34**:570-579. DOI: 10.1039/TF9383400570

[6] Nuno M, Ball RJ, Bowen CR. Photocatalytic properties of commercially available TiO_2 powders for pollution control. In: Semiconductor Photocatalysis—Materials, Mechanisms and Applications. United Kingdom: IntechOpen; 2016. pp. 613-633. DOI: 10.5772/62894

[7] Serpone N, Emeline AV. Semiconductor photocatalysis—Past, present, and future outlook. Journal of Physical Chemistry Letters. 2012;**3**: 673-677. DOI: 10.1021/jz300071j

[8] Hoffmann MR, Martin ST, Choi W, Bahnemann DW. Environmental applications of semiconductor photocatalysis. Chemical Reviews. 1995; **95**:69-96. DOI: 10.1021/cr00033a004

[9] Kisch H. Semiconductor photocatalysis—Mechanistic and synthetic aspects. Angewandte Chemie, International Edition. 2013;**52**:812-847. DOI: 10.1002/anie.201201200

[10] Herrmann J. Titania-based true heterogeneous photocatalysis. Environmental Science and Pollution Research. 2012;**19**:3655-3665. DOI: 10.1007/s11356-011-0697-8

[11] Herrmann J. Heterogeneous photocatalysis: Fundamentals and applications to the removal of various types of aqueous pollutants. Catalysis Today. 1999;**53**:115-129

[12] Pelaez M, Nolan NT, Pillai SC, Seery MK, Falaras P, Kontos AG, et al. A review on the visible light active titanium dioxide photocatalysts for environmental applications. Applied Catalysis B: Environmental. 2012;**125**: 331-349. DOI: 10.1016/j.apcatb.2012.05.036

[13] Agrawal A. Chapter 4 type I and type II band alignments in ZnO/MgZnO bilayer films. Applied Physics Letters. 2014;**105**:081603

[14] Kroemer H. Band Diagrams in Real Space: I2009. pp. 1-13

[15] Wang Y, Wang Q. Visible light driven type II heterostructures and their enhanced photocatalysis properties: A review nanoscale. Nanoscale. 2015;**5**:8326-8339. DOI: 10.1039/c3nr01577g

[16] Opoku F, Govender KK, van Sittert CGCE, Govender PP. Recent progress in the development of semiconductor-based photocatalyst materials for applications in photocatalytic water splitting and degradation of pollutants. Advanced Sustainable Systems. 2017;**1**: 1700006. DOI: 10.1002/adsu.201700006

[17] Wang Y, Wang Y, Li Y, Shi H, Xu Y, Qin H, et al. Simple synthesis of Zr-doped graphitic carbon nitride towards enhanced photocatalytic performance under simulated solar light irradiation. Catalysis Communications. 2015;**72**:24-28. DOI: 10.1016/j.catcom.2015.08.022

[18] Ibhadon AO, Fitzpatrick P. Heterogeneous photocatalysis: Recent advances and applications. 2013;**3**: 189-218. DOI: 10.3390/catal3010189

[19] Fragalà ME, Di Mauro A, Cristaldi DA, Cantarella M, Impellizzeri G, Privitera V. ZnO nanorods grown on ultrathin ZnO seed layers: Application in water treatment. Journal of Photochemistry and Photobiology A: Chemistry. 2017;**332**: 497-504. DOI: 10.1016/j.jphotochem.2016.09.032

[20] Martyna M, Diak M, Kozak M, Adriana Zaleska-Medynska EG. Some unitary, binary, and ternary non-TiO_2 photocatalysts. In: Semiconductor Photocatalysis—Materials, Mechanisms and Applications. United Kingdom: IntechOpen; 2016. DOI: 10.5772/62583

[21] Gaya UI. Heterogenous Photocatalysis Using Inorganic Semiconductor Solids. Switzerland: Springer; 2014

[22] Medina-Ramírez AH-RI. Photocatalytic Semiconductors. Switzerland: Springer; 2015. DOI: 10.1007/978-3-319-10999-2

[23] Turchi CS. Photocatalytic degradation of organic water contaminants: Mechanisms involving hydroxyl radical attack. Journal of Catalysis. 1990;**192**:178-192

[24] Ray MB. Photocatalytic oxidation of paracetamol: Dominant reactants, intermediates, and reaction mechanisms. Environmental Science & Technology. 2009;**43**:460-465

[25] Wang J, Tafen DN, Lewis JP, Hong Z, Manivannan A, Zhi M, et al. Origin of photocatalytic activity of nitrogen-doped TiO_2 nanobelts. JACS. 2009;**131**: 12290-12297

[26] Yongquan Quab XD. Progress, challenge and perspective of heterogeneous photocatalysts. The Royal Society of Chemistry. 2012;**42**: 2568-2580. DOI: 10.1039/c2cs35355e

[27] Marschall R, Wang L. Non-metal doping of transition metal oxides for visible-light photocatalysis. Catalysis Today. 2014;**225**:111-135. DOI: 10.1016/j.cattod.2013.10.088

[28] Livraghi S, Paganini MC, Giamello E, Selloni A, Di Valentin C, Pacchioni G. Origin of photoactivity of nitrogen-doped titanium dioxide under visible light. Journal of the American Chemical Society. 2006;**128**:15666-15671. DOI: 10.1021/ja064164c

[29] Sato S, Nakamura R, Abe S. Visible-light sensitization of TiO_2 photocatalysts by wet-method N doping. Applied Catalysis A: General. 2005;**284**:131-137. DOI: 10.1016/j.apcata.2005.01.028

[30] Asahi AR, Morikawa T, Ohwaki T, Aoki K, Taga Y, Lee V, et al. Visible-light photocatalysis in nitrogen-doped titanium oxides. Science. 2001;**293**: 269-271

[31] Abadias G, Paumier F, Eyidi D, Gu P, Girardeau T, Wiley J. Structure and properties of nitrogen-doped titanium dioxide thin films produced by reactive magnetron sputtering. Surface and Interface Analysis. 2010;**42**:970-973. DOI: 10.1002/sia.3220

[32] Martínez-ferrero BE, Sakatani Y, Boissière C, Grosso D, Fuertes A. Nanostructured titanium oxynitride porous thin films as efficient visible-active photocatalysts. Advanced Functional Materials. 2007;**17**:

3348-3354. DOI: 10.1002/adfm.200700396

[33] Premkumar J. Development of super-hydrophilicity on nitrogen-doped TiO_2 thin film surface by photoelectrochemical method under. Chemistry of Materials. 2004;**16**: 3980-3981. DOI: 10.1021/cm049055g

[34] Choi H, Antoniou MG, Pelaez M, De la Cruz AA, Shoemaker JA, Dionysiou DD. Mesoporous nitrogen-doped TiO_2 for the photocatalytic destruction of the cyanobacterial toxin microcystin-LR under visible light irradiation. Environmental Science & Technology. 2007;**41**:7530-7535. DOI: 10.1021/es0709122

[35] Yang J, Chen DX, Deng AP, Huang YP, Chen CC. Visible-light-driven photocatalytic degradation of microcystin-LR by Bi-doped TiO_2. Research on Chemical Intermediates. 2011;**37**:47-60. DOI: 10.1007/s11164-010-0224-4

[36] Vitiello RP, Macak JM, Ghicov A, Tsuchiya H, Dick LFP, Schmuki P. N-doping of anodic TiO_2 nanotubes using heat treatment in ammonia. Electrochemistry Communications. 2006;**8**:544-548. DOI: 10.1016/j.elecom.2006. 01.023

[37] Zong X, Sun C, Yu H, Chen ZG, Xing Z, Ye D, et al. Activation of photocatalytic water oxidation on N-doped ZnO bundle-like nanoparticles under visible light. The Journal of Physical Chemistry. 2013;**117**:4937-4942. DOI: 10.1021/jp311729b

[38] Zhang D, Gong J, Han G, Tong Z. A facile method for synthesis of N-doped ZnO mesoporous nanospheres and enhanced photocatalytic activity. Dalton Transactions. 2013;**42**:16556-16561. DOI: 10.1039/c3dt52039k

[39] Kumar R, Abdel-wahab MS, Barakat MA, Rashid J, Salah N, Al-ghamdi AA. Role of N doping on the structural, optical and photocatalytic properties of the silver deposited ZnO thin films. Journal of the Taiwan Institute of Chemical Engineers. 2016; **69**:131-138. DOI: 10.1016/j.jtice.2016.09.027

[40] Tuomisto F, Rauch C, Wagner MR, Hoffmann A, Eisermann S, Meyer BK, et al. Nitrogen and vacancy clusters in ZnO. Journal of Materials Research. 2013;**28**:1977-1983. DOI: 10.1557/jmr.2013.195

[41] Barnes TM, Olson K, Wolden CA. On the formation and stability of type conductivity in nitrogen-doped zinc oxide. Applied Physics Letters. 2005;**86**: 112. DOI: 10.1063/1.1884747

[42] Tuzemen ES, Kara K, Elagoz S, Takci DK, Altuntas I, Esen R. Structural and electrical properties of nitrogen-doped ZnO thin films. Applied Surface Science. 2014;**318**:157-163. DOI: 10.1016/j.apsusc.2014.02.118

[43] Di Valentin C, Pacchioni G, Selloni A. Theory of carbon doping of titanium dioxide. Chemistry of Materials. 2005; **17**:6656-6665. DOI: 10.1021/cm051921h

[44] Dong F, Guo S, Wang H, Li X, Wu Z. Enhancement of the visible light photocatalytic activity of C-doped TiO_2 nanomaterials prepared by a green synthetic approach. The Journal of Physical Chemistry. 2011;**115**: 13285-13292. DOI: 10.1021/jp111916q

[45] Cong Y, Li X, Qin Y, Dong Z, Yuan G, Cui Z, et al. Carbon-doped TiO_2 coating on multiwalled carbon nanotubes with higher visible light photocatalytic activity. Applied Catalysis B: Environmental. 2011;**107**: 128-134. DOI: 10.1016/j.apcatb.2011.07.005

[46] Gamage McEvoy J, Cui W, Zhang Z. Degradative and disinfective properties of carbon-doped anatase-rutile TiO_2

mixtures under visible light irradiation. Catalysis Today. 2013;**207**:191-199. DOI: 10.1016/j.cattod.2012.04.015

[47] Rockafellow EM, Stewart LK, Jenks WS. Is sulfur-doped TiO_2 an effective visible light photocatalyst for remediation? Applied Catalysis B: Environmental. 2009;**91**:554-562. DOI: 10.1016/j.apcatb.2009.06.027

[48] Umebayashi T, Yamaki T, Itoh H, Asai K. Band gap narrowing of titanium dioxide by sulfur doping. Applied Physics Letters. 2002;**454**:3-6. DOI: 10.1063/1.1493647

[49] Xu H, Zhou Y-N, Lu F, Fu Z-W. Electrochemistry of $ZnO_{1-x}S_x$ thin film with lithium. Journal of the Electrochemical Society. 2011;**158**:A285. DOI: 10.1149/1.3532037

[50] Meyer BK, Polity A, Farangis B, He Y, Hasselkamp D, Krämer T, et al. Structural properties and bandgap bowing of thin films deposited by reactive sputtering. Applied Physics Letters. 2004;**4929**:2002-2005. DOI: 10.1063/1.1825053

[51] Ratheesh R, Thankalekshmi ACR. Structure and optical band gap of $ZnO_{1-x}S_x$ thin films synthesized by chemical spray pyrolysis for application in solar cells. Journal of Applied Physics. 2012;**112**:063708. DOI: 10.1063/1.4754014

[52] Patil AB, Patil KR, Pardeshi SK. Ecofriendly synthesis and solar photocatalytic activity of S-doped ZnO. Journal of Hazardous Materials. 2010;**183**:315-323. DOI: 10.1016/j.jhazmat.2010.07.026

[53] Zheng YZ, Tao X, Hou Q, Wang DT, Zhou WL, Chen JF. Iodine-doped ZnO nanocrystalline aggregates for improved dye-sensitized solar cells. Chemistry of Materials. 2011;**23**:3-5. DOI: 10.1021/cm101525p

[54] Iwase M, Yamada K, Kurisaki T, Prieto-mahaney OO, Ohtani B, Wakita H. Applied catalysis B: Environmental visible-light photocatalysis with phosphorus-doped titanium (IV) oxide particles prepared using a phosphide compound. Applied Catalysis B: Environmental. 2013;**132–133**:39-44. DOI: 10.1016/j.apcatb.2012.11.014

[55] Mu X, Hou T, Zhang L, Jiang Y, Guan J, Chen X. Facile synthesis of phosphorus doped graphitic carbon nitride polymers with enhanced visible-light photocatalytic activity. Materials Research Bulletin. 2013;**48**:3485-3491. DOI: 10.1016/j.materresbull.2013.05.040

[56] Chai B, Yan J, Wang C, Ren Z, Zhu Y. Enhanced visible light photocatalytic degradation of Rhodamine B over phosphorus doped graphitic carbon nitride. Applied Surface Science. 2017;**391**:376-383. DOI: 10.1016/j.apsusc.2016.06.180

[57] Fan Q, Liu J, Yu Y, Zuo S, Li B. A simple fabrication for sulfur doped graphitic carbon nitride porous rods with excellent photocatalytic activity degrading RhB dye. Applied Surface Science. 2017;**391**:360-368. DOI: 10.1016/j.apsusc.2016.04.055

[58] Yan SC, Li ZS, Zou ZG. Photodegradation of rhodamine B and methyl Orange over boron-doped g-C_3N_4 under visible light irradiation. Langmuir. 2010;**26**:3894-3901. DOI: 10.1021/la904023j

[59] Dong G, Zhao K, Zhang L. Carbon self-doping induced high electronic conductivity and photoreactivity of g-C_3N_4. Chemical Communications. 2012;**48**:6178-6180. DOI: 10.1039/c2cc32181e

[60] Gole JL, Stout JD, Burda C, Lou Y, Chen X. Highly efficient formation of visible light tunable $TiO_{2-x}N_x$ photocatalysts and their transformation

at the nanoscale. The Journal of Physical Chemistry. B. 2004;**108**:1230-1240

[61] Jiang W, Ullah N, Divitini G, Ducati C, Kumar RV, Ding Y, et al. Vertically oriented TiO_xN_y nanopillar arrays with embedded Ag nanoparticles for visible-light photocatalysis. Langmuir. 2012;**28**:5427-5431. DOI: 10.1021/la203617u

[62] Rhee CH, Bae SW, Lee JS. Template-free hydrothermal synthesis of high surface area nitrogen-doped Titania photocatalyst active under visible light. Chemistry Letters. 2005;**34**:660-661. DOI: 10.1246/cl.2005.660

[63] Livraghi S, Pelaez M, Biedrzycki J, Corazzari I, Giamello E, Dionysiou DD. Influence of the chemical synthesis on the physicochemical properties of N-TiO_2 nanoparticles. Catalysis Today. 2013;**209**:54-59. DOI: 10.1016/j.cattod.2012.12.020

[64] Fang J, Shi F, Bu J, Ding J, Xu S, Bao J, et al. One-step synthesis of bifunctional TiO_2 catalysts and their photocatalytic activity. Journal of Physical Chemistry C. 2010;**114**: 7940-7948

[65] Cong Y, Zhang J, Chen F, Anpo M. Synthesis and characterization of nitrogen-doped TiO_2 nanophotocatalyst with high visible light activity. Journal of Physical Chemistry C. 2007;**111**: 6976-6982. DOI: 10.1021/jp0685030

[66] Lin L, Zheng RY, Xie JL, Zhu YX, Xie YC. Synthesis and characterization of phosphor and nitrogen co-doped titania. Applied Catalysis B: Environmental. 2007;**76**:196-202. DOI: 10.1016/j.apcatb.2007.05.023

[67] Li F, Yin X. Investigation on F-B-S tri-doped nano-TiO_2 films for the photocatalytic degradation of organic dyes. Journal of Nanoparticle Research. 2011;**13**:4839-4846. DOI: 10.1007/s11051-011-0461-5

[68] Zhao W, Ma W, Chen C, Zhao J, Shuai Z. Efficient degradation of toxic organic pollutants with $Ni_2O_3/TiO_{2-x}B_x$ under visible irradiation. Journal of the American Chemical Society. 2004;**126**: 4782-4783. DOI: 10.1021/ja0396753

[69] Dong F, Zhao W, Wu Z. Characterization and photocatalytic activities of C, N and S co-doped TiO_2 with 1D nanostructure prepared by the nano-confinement effect. Nanotechnology. 2008;**19**:10. DOI: 10.1088/0957-4484/19/36/365607

[70] Ohno T, Tsubota T, Toyofuku M, Inaba R. Photocatalytic activity of a TiO_2 photocatalyst doped with C^{4+} and S^{4+} ions having a rutile phase under visible light. 2004;**98**:255-258

[71] Hamal DB, Haggstrom JA, Marchin GL, Ikenberry MA, Hohn K, Klabunde KJ. A multifunctional biocide/sporocide and photocatalyst based on titanium dioxide (TiO_2) Codoped with silver, carbon, and sulfur. Langmuir. 2010;**26**: 2805-2810. DOI: 10.1021/la902844r

[72] Pelaez M, Falaras P, Likodimos V, Kontos AG, De AA, Kevin O, et al. Synthesis, structural characterization and evaluation of sol-gel-based NF-TiO_2 films with visible light-photoactivation for the removal of microcystin-LR. Applied Catalysis B: Environmental. 2010;**99**:378-387. DOI: 10.1016/j.apcatb.2010.06.017

[73] Huang D, Liao S, Liu J, Dang Z, Petrik L. Preparation of visible-light responsive N–F-codoped TiO_2 photocatalyst by a sol-gel-solvothermal method. Journal of Photochemistry and Photobiology A: Chemistry. 2006;**184**: 282-288. DOI: 10.1016/j.jphotochem.2006.04.041

[74] Lv Y, Fu Z, Yang B, Xu J, Wu M, Zhu C, et al. Preparation N–F-codoped

TiO_2 nanorod array by liquid phase deposition as visible light photocatalyst. Materials Research Bulletin. 2011;**46**: 361-365. DOI: 10.1016/j.materresbull.2010.12.011

[75] Ao Y, Xu J, Fu D, Yuan C. Synthesis of C, N, S-tridoped mesoporous titania with enhanced visible light-induced photocatalytic activity. Microporous and Mesoporous Materials. 2009;**122**: 1-6. DOI: 10.1016/j.micromeso.2008.11.010

[76] Wang P, Yap P, Lim T. C-N-S tridoped TiO_2 for photocatalytic degradation of tetracycline under visible-light irradiation. Applied Catalysis A: General. 2011;**399**:252-261. DOI: 10.1016/j.apcata.2011.04.008

[77] Ohno T, Tsubota T, Nakamura Y, Sayama K. Preparation of S, C cation-codoped $SrTiO_3$ and its photocatalytic activity under visible light. Applied Catalysis A: General. 2005;**288**:74-79. DOI: 10.1016/j.apcata.2005.04.035

[78] Zou F, Jiang Z, Qin X, Zhao Y, Jiang L, Zhi J. Template-free synthesis of mesoporous N-doped $SrTiO_3$ perovskite with high visible-light-driven photocatalytic activity. Chemical Communications. 2012;**48**:8514-8516. DOI: 10.1039/c2cc33797e

[79] Han P, Su Y, Meng Y, Wang S, Jia Q, Wang X. Preparation and photocatalytic performance of iodine-doped $NaTaO_3$ nanoparticles. Journal of Nanoscience and Nanotechnology. 2011;**11**: 9600-9606. DOI: 10.1166/jnn.2011.5277

[80] Fu H, Zhang S, Zhang L, Zhu Y. Visible-light-driven $NaTaO_{3-x}N_x$ catalyst prepared by a hydrothermal process. Materials Research Bulletin. 2008;**43**:864-872. DOI: 10.1016/j.materresbull.2007.05.013

[81] Jiang L, Wang Q, Li C, Yuan J, Shangguan W. ZrW_2O_8 photocatalyst and its visible-light sensitization via sulfur anion doping for water splitting. International Journal of Hydrogen Energy. 2010;**35**:7043-7050. DOI: 10.1016/j.ijhydene.2009.12.187

[82] Zhang J, Huang Z, Xu Y, Kang F. Hydrothermal synthesis of iodine-doped Bi_2WO_6 nanoplates with enhanced visible and ultraviolet-induced photocatalytic activities. International Journal of Photoenergy. 2012;**2012**. DOI: 10.1155/2012/915386

[83] Zhang Z, Wang W, Wang L, Sun S. Enhancement of visible-light photocatalysis by coupling with narrow-band-gap semiconductor: A case study on Bi_2S_3/Bi_2WO_6. Applied Materials & Interfaces. 2012;**4**:593-597. DOI: 10.1021/am2017199

[84] Shi R, Huang G, Lin J, Zhu Y. Photocatalytic activity enhancement for Bi_2WO_6 by fluorine substitution. Journal of Physical Chemistry C. 2009;**113**: 19633-19638

[85] Shifu C, Wei Z, Sujuan Z, Wei L. Preparation, characterization and photocatalytic activity of N-containing ZnO powder. Chemical Engineering Journal. 2009;**148**:263-269. DOI: 10.1016/j.cej.2008.08.039

[86] Li X, Pan H, Hu Q, Zhang C. Nitrogen-doped SiO_2–HNb_3O_8 for rhodamine B photodegradation under visible light. Journal of Alloys and Compounds. 2011;**509**:6252-6256. DOI: 10.1016/j.jallcom.2011.03.033

[87] Takeuchi M, Shimizu Y, Yamagawa H, Nakamuro T, Anpo M. Preparation of the visible light responsive N_3-doped WO_3 photocatalyst by a thermal decomposition of ammonium paratungstate. Applied Catalysis B: Environmental. 2011;**110**:1-5. DOI: 10.1016/j.apcatb.2011.08.004

[88] Liu Y, Li Y, Li W, Han S, Liu C. Photoelectrochemical properties and photocatalytic activity of

nitrogen-doped nanoporous WO_3 photoelectrodes under visible light. Applied Surface Science. 2012;**258**: 5038-5045. DOI: 10.1016/j.apsusc. 2012.01.080

[89] Liu S, Yin K, Ren W, Cheng B, Yu J. Tandem photocatalytic oxidation of Rhodamine B over surface fluorinated bismuth vanadate crystals. Journal of Materials Chemistry. 2012;**22**: 17759-17767. DOI: 10.1039/c2jm33337f

[90] Ji K, Jiang H, Liu Y, Deng J, Dai H, Zhang L. Porous F-doped $BiVO_4$: Synthesis and enhanced photocatalytic performance for the degradation of phenol under visible-light illumination. Solid State Sciences. 2013;**17**:2127. DOI: 10.1016/j. solidstatesciences.2012.12.009

[91] Zhao Z, Dai H, Deng J, Liu Y, Au CT. Enhanced visible-light photocatalytic activities of porous olive-shaped sulfur-doped $BiVO_4$-supported cobalt oxides. Solid State Sciences. 2013;**18**:98-104. DOI: 10.1016/j.solidstatesciences.2013.01.009

[92] Jia L, Li J, Fang W, Song H, Li Q, Tang Y. Visible-light-induced photocatalyst based on C-doped $LaCoO_3$ synthesized by novel microorganism chelate method. Catalysis Communications. 2009;**10**:1230-1234. DOI: 10.1016/j.catcom.2009. 01.025

[93] Martha S, Parida KM. Fabrication of nano N-doped $In_2Ga_2ZnO_7$ for photocatalytic hydrogen production under visible light. International Journal of Hydrogen Energy. 2012;**37**: 17936-17946. DOI: 10.1016/j.ijhydene. 2012.09.025

[94] Gasparotto A, Barreca D, Bekermann D, Devi A, Fischer R a, Fornasiero P, et al. F-doped Co_3O_4 photocatalysts for sustainable H_2 generation from water/ethnol. Journal of the American Chemical Society. 2011;**133**:193-625. DOI: 10.1021/ ja210078d

[95] Sreethawong T, Ngamsinlapasathian S, Yoshikawa S. Synthesis of crystalline mesoporous-assembled ZrO_2 nanoparticles via a facile surfactant-aided sol-gel process and their photocatalytic dye degradation activity. Chemical Engineering Journal. 2013; **228**:256-262. DOI: 10.1016/j.cej.2013. 04.111

[96] Dolat D, Quici N, Kusiak-nejman E, Morawski AW, Puma GL. Environmental one-step, hydrothermal synthesis of nitrogen, carbon co-doped titanium dioxide (N, C TiO_2) photocatalysts. Effect of alcohol degree and chain length as carbon dopant precursors on photocatalytic activity and cata. Applied Catalysis B: Environmental. 2012;**115–116**:81-89. DOI: 10.1016/j.apcatb.2011.12.007

[97] Sun H, Bai Y, Cheng Y, Jin W, Xu N. Preparation and characterization of visible-light-driven carbon-sulfur-codoped TiO_2 photocatalysts. Industrial and Engineering Chemistry Research. 2006;**45**:4971-4976. DOI: 10.1021/ ie060350f

[98] Uk H, Chang S, Hae S, Son B, Jae S, Jin H, et al. Highly visible-light active nanoporous TiO_2 photocatalysts for efficient solar photocatalytic applications. Applied Catalysis B: Environmental. 2013;**129**:106-113. DOI: 10.1016/j.apcatb.2012.09.010

[99] Yang K, Dai Y, Huang B, Whangbo MH. Density functional characterization of the band edges, the band gap states, and the preferred doping sites of halogen-doped TiO_2. Chemistry of Materials. 2008;**20**:6528-6534. DOI: 10.1021/cm801741m

[100] Zhang L, Zhang H, Xu H, Deng F, Zheng Z. Hierarchical chlorine-doped rutile TiO_2 spherical clusters of

nanorods: Large-scale synthesis and high photocatalytic activity. Journal of Solid State Chemistry. 2008;**181**: 2516-2522. DOI: 10.1016/j.jssc.2008.06.019

[101] Luo H, Takata T, Lee Y, Zhao J, Domen K, Yan Y. Photocatalytic activity enhancing for titanium dioxide by Co-doping with bromine and chlorine. Chemistry of Materials. 2004;**16**: 846-849. DOI: 10.1021/cm035090w

[102] Andersen J, Han C, O'Shea K, Dionysiou DD. Revealing the degradation intermediates and pathways of visible light-induced NF-TiO_2 photocatalysis of microcystin-LR. Applied Catalysis B: Environmental. 2014;**154–155**:259-266. DOI: 0.1016/j.apcatb.2014.02.025

[103] Nisar J, Wang B, Moyses C, Ferreira A. Band gap engineering by anion doping in the photocatalyst $BiTaO_4$: First principle calculations. International Journal of Hydrogen Energy. 2012;**37**:3014-3018. DOI: 10.1016/j.ijhydene.2011.11.068

[104] Yang K, Huang W, Xu L, Luo K, Yang Y, Huang G. Materials science in semiconductor processing insights into enhanced visible-light photocatalytic activity of CeO_2 doped with nonmetal impurity from the first principles. Materials Science in Semiconductor Processing. 2016;**41**:200-208. DOI: 10.1016/j.mssp.2015.08.039

[105] Bagwasi S, Tian B, Zhang J, Nasir M. Synthesis, characterization and application of bismuth and boron Co-doped TiO_2: A visible light active photocatalyst. Chemical Engineering Journal. 2013;**217**:108-118. DOI: 10.1016/j.cej.2012.11.080

[106] Di Valentin C, Pacchioni G. Trends in non-metal doping of anatase TiO_2: B, C, N and F. Catalysis Today. 2013;**206**: 12-18. DOI: 10.1016/j.cattod.2011.11.030

[107] Wang G, Chen H, Wu G, Kuang A. Hybrid density functional study on mono- and Co-doped $NaNbO_3$ for visible-light photocatalysis. ChemPhysChem. 2016;**17**:489-499. DOI: 10.1002/cphc.201501037

[108] Murakami N, Chiyoya T, Tsubota T, Ohno T. Switching redox site of photocatalytic reaction on titanium(IV) oxide particles modified with transition-metal ion controlled by irradiation wavelength. Applied Catalysis A: General. 2008;**348**:148-152. DOI: 10.1016/j.apcata.2008.06.040

[109] Carneiroa JO, Portinha VT, Dupak L, Magalhaes A, Coutinho P. Study of the deposition parameters and Fe-dopant effect in the photocatalytic activity of TiO_2 films prepared by dc reactive magnetron sputtering. Vacuum. 2005;**78**:37-46. DOI: 10.1016/j.vacuum.2004.12.012

[110] Wu JCS, Chen CH. A visible-light response vanadium-doped titania nanocatalyst by sol-gel method. Journal of Photochemistry and Photobiology A: Chemistry. 2004;**163**:509-515. DOI: 10.1016/j.jphotochem.2004.02.007

[111] Rimoldi L, Pargoletti E, Meroni D, Falletta E, Cerrato G, Turco F, et al. Concurrent role of metal (Sn, Zn) and N species in enhancing the photocatalytic activity of TiO_2 under solar light. Catalysis Today. 2018;**313**:4046. DOI: 10.1016/j.cattod.2017.12.017

[112] Rashid J, Barakat MA, Salah N, Habib SS. Ag/ZnO nanoparticles thin films as visible light photocatalysts. The Royal Society of Chemistry. 2014;**4**: 56892-56899. DOI: 10.1039/c4ra12990c

[113] Yayapao O, Thongtem T, Phuruangrat A, Thongtem S. Ultrasonic-assisted synthesis of Nd-doped ZnO for photocatalysis. Materials Letters. 2013; **90**:83-86. DOI: 10.1016/j.matlet.2012.09.027

[114] Phuruangrat A, Yayapao O, Thongtem T, Thongtem S. Preparation, characterization and photocatalytic properties of Ho doped ZnO nanostructures synthesized by sonochemical method. Superlattices and Microstructures. 2014;**67**:118-126. DOI: 10.1016/j.spmi.2013.12.023

[115] Khatamian M, Khandar AA, Divband B, Haghighi M, Ebrahimiasl S. Heterogeneous photocatalytic degradation of 4-nitrophenol in aqueous suspension by Ln (La^{3+}, Nd^{3+} or Sm^{3+}) doped ZnO nanoparticles. Journal of Molecular Catalysis A: Chemical. 2012; **365**:120-127. DOI: 10.1016/j.molcata.2012.08.018

[116] Ding K, Wen L, Xu L, Wu H, Ye Y, Zhang Y. Predicting the electronic and optical properties of IB metals doped monoclinic $BiVO_4$: First principle calculations. International Journal of Quantum Chemistry. 2015;**116**:388-395. DOI: 10.1002/qua.25051

[117] Tang JW, Wang DF, Zou ZG, Je Y. Modification of photophysical properties of WO_3 by doping different metals. Materials Science Forum. 2003; **425**:163-166. DOI: 10.4028/www.scientific.net/MSF.423-425.163

[118] Xue Q, Liu Y, Zhou Q, Utsumi M, Zhang Z, Sugiura N. Photocatalytic degradation of geosmin by Pd nanoparticle modified WO_3 catalyst under simulated solar light. Chemical Engineering Journal. 2016;**283**:614-621. DOI: 10.1016/j.cej.2015.08.016

[119] Wei W, Ying D, Hao J, Huang B. Density functional characterization of the of Cr-doped $SrTiO_3$. Journal of Physics D: Applied Physics. 2009;**42**. DOI: 10.1088/0022-3727/42/5/055401

[120] Liu H, Peng T, Ke D, Peng Z, Yan C. Preparation and photocatalytic activity of dysprosium doped tungsten trioxide nanoparticles. Materials Chemistry and Physics. 2007;**104**: 377-383. DOI: 10.1016/j.matchemphys.2007.03.028

[121] Li J, Hu W, Zhong J, Zeng J, Huang S, Xiao Z, et al. Photo-induced charge separation and photocatalytic activity of Ga-doped SnO_2. Applied Physics A: Materials Science & Processing. 2014; **116**:2149-2156. DOI: 10.1007/s00339-014-8428-x

[122] Li Z, Shen Y, Guan Y, Hu Y, Lin Y, Nan CW. Bandgap engineering and enhanced interface coupling of graphene-$BiFeO_3$ nanocomposites as efficient photocatalysts under visible light. Journal of Materials Chemistry A. 2014;**2**:1967-1973. DOI: 10.1039/c3ta14269h

[123] Hu S, Li F, Fan Z, Wang F, Zhao Y, Lv Z. Band gap-tunable potassium doped graphitic carbon nitride with enhanced mineralization. Dalton Transactions. 2014;**44**:1084-1092. DOI: 10.1039/c4dt02658f

[124] Rong X, Qiu F, Rong J, Zhu X, Yan J, Yang D. Enhanced visible light photocatalytic activity of W-doped porous. Materials Letters. 2016;**164**: 127-131. DOI: 10.1016/j.matlet.2015.10.131

[125] Wang Y, Li Y, Bai X, Cai Q, Liu C, Zuo Y, et al. Facile synthesis of Y-doped graphitic carbon nitride with enhanced photocatalytic performance. Catalysis Communications. 2016;**84**:179-182. DOI: 10.1016/j.catcom.2016.06.020

[126] Song X, Tao H, Chen L, Sun Y. Synthesis of Fe/g-C_3N_4 composites with improved visible light photocatalytic activity. Materials Letters. 2014;**116**: 265-267. DOI: 10.1016/j.matlet.2013.11.043

[127] Chatterjee D, Dasgupta S. Visible light induced photocatalytic degradation of organic pollutants. Journal of Photochemistry and Photobiology C: Photochemistry Reviews. 2005;**6**:

186-205. DOI: 10.1016/j.jphotochemrev.2005.09.001

[128] Abe R, Sayama K, Arakawa H. Dye-sensitized photocatalysts for efficient hydrogen production from aqueous I— Solution under visible light irradiation. Journal of Photochemistry and Photobiology A: Chemistry. 2004;**166**: 115-122. DOI: 10.1016/j.jphotochem.2004.04.031

[129] Chu W, Chan KH, Jafvert CT, Chan YS. Removal of phenylurea herbicide monuron via riboflavin-mediated photosensitization. Chemosphere. 2007;**69**:177-183. DOI: 10.1016/j.chemosphere.2007.04.055

[130] Whitehead K, Hedges JI. Photodegradation and photosensitization of mycosporine-like amino acids. Journal of Photochemistry and Photobiology B: Biology. 2005;**80**: 115-121. DOI: 10.1016/j.jphotobiol.2005.03.008

[131] Kaur S, Singh V. Visible light induced sonophotocatalytic degradation of reactive red dye 198 using dye sensitized TiO_2. Ultrasonics Sonochemistry. 2007;**14**:531-537. DOI: 10.1016/j.ultsonch.2006.09.015

[132] Lee C, Hyeon T, Lee H. Visible light-induced degradation of carbon tetrachloride on. Environmental Science & Technology. 2001;**35**:966-970

[133] Shim M, Mcdaniel H, Oh N. Prospects for strained type-II nanorod heterostructures. The Journal of Physical Chemistry Letters. 2011;**2**: 2722-2727

[134] Tian J, Zhenhuan Zhaoa AK, Boughton RI, Liu H. Nanostructured surface heterostructures: A review. Chemical Society Reviews. 2014;**43**: 6920-6937. DOI: 10.1039/x0xx00000x

[135] Beydoun D, Amal R, Low G. Role of nanoparticles in photocatalysis. Journal of Nanoparticle Research. 1999; **1**:439-458. DOI: 10.1023/A:1010044830871

[136] Ghows N, Entezari MH. Fast and easy synthesis of core—Shell nanocrystal (CdS/TiO_2) at low temperature by micro-emulsion under ultrasound. Ultrasonics Sonochemistry. 2011;**18**:629-634. DOI: 10.1016/j.ultsonch.2010.08.003

[137] Jia H, Xu H, Hu Y, Tang Y, Zhang L. TiO_2 @ CdS core—Shell nanorods films: Fabrication and dramatically enhanced photoelectrochemical properties. Electrochemistry Communications. 2007;**9**:354-360. DOI: 10.1016/j.elecom.2006.10.010

[138] Covei M, Perniu D, Bogatu C, Duta A. CZTS-TiO_2 thin film heterostructures for advanced photocatalytic wastewater treatment. Catalysis Today. 2019;**321–322**:172-177. DOI: 10.1016/j.cattod.2017.12.003

[139] Rashid J, Barakat MA, Ruzmanova Y. $Fe_3O_4/SiO_2/TiO_2$ nanoparticles for photocatalytic degradation of 2-chlorophenol in simulated wastewater. Environmental Science and Pollution Research. 2015;**22**:3149-3157. DOI: 10.1007/s11356-014-3598-9

[140] Pan L, Shen G, Zhang J, Wei X, Wang L, Zou J, et al. TiO_2—ZnO composite sphere decorated with ZnO clusters for E ff ective charge isolation in photocatalysis. Industrial and Engineering Chemistry Research. 2015; **54**:7226-7232. DOI: 10.1021/acs.iecr.5b01471

[141] Lee G, Kang M. Physicochemical properties of core/shell structured pyrite FeS_2/anatase TiO_2 composites and their photocatalytic hydrogen production performances. Current Applied Physics. 2013;**13**:1482-1489. DOI: 10.1016/j.cap.2013.05.002

[142] Rashid J, Saleem S, Awan SU, Iqbal A, Kumar R, Barakat MA, et al.

Stabilized fabrication of anatase-TiO_2/FeS_2 (pyrite) semiconductor composite nanocrystals for enhanced solar light-mediated photocatalytic degradation of methylen blue. RSC Advances. 2018;**8**: 11935-11945. DOI: 10.1039/c8ra02077a

[143] Rashid J, Barakat M. $InVO_4$/TiO_2 composite for visible-light photocatalytic degradation of 2-chlorophenol in wastewater. Environmental Technology. 2014;**35**: 2153-2159. DOI: 10.1080/09593330.2014.895051

[144] Rashid J, Barakat MA. Ag_3PO_4 Enhanced TiO_2 for Visible Light Photocatalysis of 2-Chlorophenol in Wastewater. Institute of Research Engineers and Doctors, USA; 2014. pp. 3-7

[145] Li Y, Zhao H, Yang M. TiO_2 nanoparticles supported on PMMA nanofibers for photocatalytic degradation of methyl orange. Journal of Colloid and Interface Science. 2017;**508**: 500-507. DOI: 10.1016/j.jcis.2017.08.076

[146] Rashid J, Abbas A, Chang LC, Iqbal A, Ihsan-ul-Haq A, Rehman SUA, et al. Butterfly cluster like lamellar BiOBr/TiO_2 nanocomposite for enhanced sunlight photocatalytic mineralization of aqueous ciprofloxacin. Science of the Total Environment Elsevier B. 2019;**665**: 668-677. DOI: 10.1016/j.scitotenv.2019.02.145

[147] Jana TK, Pal A, Chatterjee K. Self assembled flower like CdS–ZnO nanocomposite and its photo catalytic activity. Journal of Alloys and Compounds. 2014;**583**:510-515. DOI: 10.1016/j.jallcom.2013.08.184

[148] Cho S, Jang JW, Kim J, Lee JS, Choi W, Lee KH. Three-dimensional type II ZnO/ZnSe heterostructures and their visible light photocatalytic activities. Langmuir. 2011;**27**:10243-10250. DOI: 10.1021/la201755w

[149] Wang Z, Huang B, Dai Y, Qin X, Zhang X, Wang P, et al. Highly photocatalytic ZnO/In_2O_3 heteronanostructures synthesized by a coprecipitation method. The Journal of Physical Chemistry. 2009;**113**:4612-4617

[150] Hamrouni A, Moussa N, Di Paola A, Parrino F, Houas A, Palmisano L. Characterization and photoactivity of coupled ZnO-$ZnWO_4$ catalysts prepared by a sol-gel method. Applied Catalysis B: Environmental. 2014;**154–155**:379-385. DOI: 10.1016/j.apcatb.2014.02.042

[151] Hernández S, Hidalgo D, Sacco A, Chiodoni A, Lamberti A, Cauda V, et al. Comparison of photocatalytic and transport properties of TiO_2 and ZnO nanostructures for solar-driven water splitting. Physical Chemistry Chemical Physics. 2015;**17**:7775-7786. DOI: 10.1039/c4cp05857g

[152] Katsumata H, Gomathisankar P, Funasaka K, Kaneco S, Suzuki T, Hachisuka K. Photocatalytic hydrogen production with CuS/ZnO from aqueous Na_2S + Na_2SO_3 solution. International Journal of Hydrogen Energy. 2013;**38**: 8625-8630. DOI: 10.1016/j.ijhydene.2013.04.131

[153] He Z, Shi Y, Gao C, Wen L, Chen J, Song S. BiOCl/$BiVO_4$ p–n heterojunction with enhanced photocatalytic activity under visible-light irradiation. The Journal of Physical Chemistry. 2014;**118**:389-398. DOI: 10.1021/jp409598s

[154] Tahir MB, Nabi G, Rafique M, Khalid NR. Nanostructured-based WO_3 photocatalysts: Recent development, activity enhancement, perspectives and applications for wastewater treatment. International journal of Environmental Science and Technology. 2017;**14**: 2519-2542. DOI: 10.1007/s13762-017-1394-z

[155] Shamaila S, Khan A, Sajjad L, Chen F, Zhang J. WO_3/BiOCl, a novel heterojunction as visible light

photocatalyst. Journal of Colloid and Interface Science. 2011;**356**:465-472. DOI: 10.1016/j.jcis.2011.01.015

[156] Huang T, Lin X, Xing J, Wang W, Shan Z, Huang F. Photocatalytic activities of hetero-junction semiconductors $WO_3/SrNb_2O_6$. Materials Science & Engineering, B: Solid-State Materials for Advanced Technology. 2007;**141**:49-54. DOI: 10.1016/j.mseb.2007.05.007

[157] Zhang J, Yu K, Yu Y, Lou LL, Yang Z, Yang J, et al. Highly effective and stable Ag_3PO_4/WO_3 photocatalysts for visible light degradation of organic dyes. Journal of Molecular Catalysis A: Chemical. 2014;**391**:12-18. DOI: 10.1016/j.molcata.2014.04.010

[158] Huang Z, Zou J, Pan L, Wang S, Zhang X, Wang L. Synergetic promotion on photoactivity and stability of $W_{18}O_{49}/TiO_2$ hybrid. Applied Catalysis B: Environmental. 2014;**147**: 167-174. DOI: 10.1016/j.apcatb.2013.08.038

[159] Aslam I, Cao C, Tanveer M, Farooq MH, Khan SW, Tahir M, et al. A novel Z-scheme $WO_3/CdWO_4$ photocatalyst with enhanced visible-light photocatalytic activity for the degradation of organic pollutants. RSC Advances. 2015;**5**:6019-6026. DOI: 10.1039/C4RA15847D

[160] Zaleska A, Winiarski M, Klimczuk T, Bajorowicz B, Cybula A. Surface properties and photocatalytic activity of $KTaO_3$, CdS, MoS_2 semiconductors and their binary and ternary semiconductor composites. Molecules. 2014;**19**: 15339-15360. DOI: 10.3390/molecules190915339

[161] Pirhashemi M, Habibi-yangjeh A. Ternary $ZnO/AgBr/Ag_2CrO_4$ nanocomposites with tandem n–n heterojunctions as novel visible-light-driven photocatalysts with excellent activity. Ceramics International. 2015; **41**:14383-14393. DOI: 10.1016/j.ceramint.2015.07.072

[162] Zhang W, Li Y, Song Q, Zhang J, Chen X, Liu D, et al. Multi-pathway photoelectron migration in globular flower-like $In_2O_3/AgBr/Bi_2WO_6$ synthesized by microwave-assisted method with enhanced photocatalytic activity. Journal of Molecular Catalysis A: Chemical. 2015;**414**:27-36. DOI: 10.1016/j.molcata.2015.12.023

[163] Mohamed MM, Ahmed SA, Khairou KS. Unprecedented high photocatalytic activity of nanocrystalline $WO_3/NiWO_4$ hetero-junction towards dye degradation: Effect of template and synthesis conditions. Applied Catalysis B: Environmental. 2014;**150–151**:63-73. DOI: 10.1016/j.apcatb.2013.12.001

[164] Xu YS, De Zhang W. Monodispersed Ag_3PO_4 nanocrystals loaded on the surface of spherical Bi_2MoO_6 with enhanced photocatalytic performance. Dalton Transactions. 2013;**42**:1094-1101. DOI: 10.1039/c2dt31634j

[165] Cao J, Luo B, Lin H, Xu B, Chen S. Visible light photocatalytic activity enhancement and mechanism of AgBr/Ag_3PO_4 hybrids for degradation of methyl orange. Journal of Hazardous Materials. 2012;**217–218**:107-115. DOI: 10.1016/j.jhazmat.2012.03.002

[166] Feng Y, Yan X, Liu C, Hong Y, Zhu L, Zhou M, et al. Applied surface science hydrothermal synthesis of CdS/Bi_2MoO_6 heterojunction photocatalysts with excellent visible-light-driven photocatalytic performance. Applied Surface Science. 2015;**353**:87-94. DOI: 10.1016/j.apsusc.2015.06.061

[167] Ghugal SG, Umare SS, Sasikala R. Photocatalytic mineralization of anionic dyes using bismuth doped CdS–Ta_2O_5 composite. RSC Advances. 2015;**5**:

63393-63400. DOI: 10.1039/C5RA09974A

[168] Kong L, Jiang Z, Xiao T, Lu L, Jones O, Edwards PP. Exceptional visible-light-driven photocatalytic activity over. Chemical Communications. 2011;**47**: 5512-5514. DOI: 10.1039/c1cc10446b

[169] Liu C, Li P, Wu G, Luo B, Lin S, Ren A, et al. Enhanced photoelectrochemical and photocatalytic activity by $Cu_2O/SrTiO_3$ p–n heterojunction via a facile deposition–precipitation technique. RSC Advances. 2015;**5**:33938-33945. DOI: 10.1039/C5RA03086B

[170] Hong E, Kim D, Kim JH. Heterostructured metal sulfide (ZnS-CuS-CdS) photocatalyst for high electron utilization in hydrogen production from solar water splitting. Journal of Industrial and Engineering Chemistry. 2014;**20**:3869-3874. DOI: 10.1016/j.jiec.2013.12.092

[171] Moussa N, Houas A, Parrino F, Di Paola A, Hamrouni A, Palmisano L. Photocatalytic activity of binary and ternary SnO_2–ZnO–$ZnWO_4$ nanocomposites. Journal of Photochemistry and Photobiology A: Chemistry. 2015;**309**:47-54. DOI: 10.1016/j.jphotochem.2015.05.001

[172] Zhao Z, Yanjuan S, Dong F. Graphitic carbon nitride based nanocomposites: A review. Nanoscale. 2015;7:15-37. DOI: 10.1039/c4nr03008g

[173] Song G, Chu Z, Jin W, Sun H. Enhanced performance of g-C_3N_4/TiO_2 photocatalysts for degradation of organic pollutants under visible light. Chinese Journal of Chemical Engineering. 2015;**23**:1326-1334. DOI: 10.1016/j.cjche.2015.05.003

[174] Lv J, Dai K, Zhang J, Geng L, Liang C, Liu Q, et al. Applied surface science facile synthesis of Z-scheme graphitic-C_3N_4/Bi_2MoO_6 nanocomposite for enhanced visible photocatalytic properties. Applied Surface Science. 2015;**358**:377-384. DOI: 10.1016/j.apsusc.2015.06.183

[175] Dai S, Hua F, Li W, Hou H, Yue J, Feng C. Fabrication of sulfur-doped g-C_3N_4/Au/CdS Z-scheme photocatalyst to improve the photocatalytic performance under visible light. Applied Catalysis B: Environmental. 2015;**168–169**:465-471. DOI: 10.1016/j.apcatb.2015.01.012

[176] Fu J, Chang B, Yanlong Tian FX, Dong X. Novel C_3N_4-CdS composite photocatalysts with organic-inorganic heterojunctions: In suit synthesis, exceptional activity, high stability and photocatalytic mechanism. Journal of Materials Chemistry A. 2013;**1**:3083-3090. DOI: 10.1039/c2ta00672c

[177] Tong Z, Yang D, Shi J, Nan Y, Sun Y, Jiang Z, et al. Three-dimensional porous aerogel constructed by g-C_3N_4 and graphene oxide nanosheets with excellent visible-light photocatalytic performance. Applied Materials & Interfaces. 2015;7:25693-25701. DOI: 10.1021/acsami.5b09503

[178] Han Q, Chen N, Zhang J, Qu L. Graphene/graphitic carbon nitride hybrids for catalysis. Materials Horizons. 2017:832-850. DOI: 10.1039/c7mh00379j

[179] Mousavi M, Habibi-Yangjeh A. Integration of $NiWO_4$ and Fe_3O_4 with graphitic carbon nitride to fabricate novel magnetically recoverable visible-light-driven photocatalysts. Journal of Materials Science. 2018;**53**:9046-9063. DOI: 10.1007/s10853-018-2213-8

[180] Zhang S, Li J, Zeng M, Zhao G, Xu J, Hu W, et al. In situ synthesis of water-soluble magnetic graphitic carbon nitride photocatalyst and its synergistic catalytic performance. ACS Applied Materials & Interfaces. 2013;**5**:12735-12743

[181] Zou L, Wang H, Yuan G, Wang X. Magnetically separable CdS/$ZnFe_2O_4$ composites with highly efficient photocatalytic activity and photostability under visible light. ACS Applied Nano Materials. 2018;**1**:831-838. DOI: 10.1021/acsanm.7b00243

[182] Chai B, Liao X, Song F, Zhou H. Fullerene modified C_3N_4 composites with enhanced photocatalytic activity under visible light irradiation. Dalton Transactions. 2014;**43**:982-989. DOI: 10.1039/c3dt52454j

[183] Fu H, Xu T, Zhu S, Zhu Y. Photocorrosion inhibition and enhancement of photocatalytic activity for ZnO via hybridization with C_6O. Environmental Science & Technology. 2008;**42**:8064-8069. DOI: 10.1021/es801484x

[184] Zhu Y, Bai X, Yao W, Wang L, Wang Y. Enhanced oxidation ability of g-C_3N_4 photocatalyst via C60 modification. Applied Catalysis B: Environmental. 2014b;**152–153**:262-270. DOI: 10.1016/j.apcatb.2014.01.046

[185] Zhu H, Chen D, Yue D. In-situ synthesis of g-C_3N_4-$P_{25}TiO_2$ composite with enhanced visible light photoactivity. Journal of Nanoparticle Research. 2014;**16**:1-10. DOI: 10.1007/s11051-014-2632-7

[186] Huang L, Xu H, Li Y, Li H, Cheng X, Xia J, et al. Visible-light-induced WO_3/g-C_3N_4 composites with enhanced photocatalytic activity. Dalton Transactions. 2013;**4**:8606-8616. DOI: 10.1039/c3dt00115f

[187] Tian N, Huang H, Zhang Y. Mixed-calcination synthesis of $CdWO_4$/g-C_3N_4 heterojunction with enhanced visible-light-driven photocatalytic activity. Applied Surface Science. 2015;**358**:343-349. DOI: 10.1016/j.apsusc.2015.07.154

[188] Xing C, Wu Z, Jiang D, Chen M. Hydrothermal synthesis of In_2S_3/g-C_3N_4 heterojunctions with enhanced photocatalytic activity. Journal of Colloid and Interface Science. 2014;**433**:9-15. DOI: 10.1016/j.jcis.2014.07.015

[189] Liu Q, Fan C, Tang H, Sun X, Yang J, Cheng X. One-pot synthesis of g-C_3N_4/V_2O_5 composites for visible light-driven photocatalytic activity. Applied Surface Science. 2015;**358**:188-195. DOI: 10.1016/j.apsusc.2015.09.010

[190] Sun JX, Yuan YP, Qiu LG, Jiang X, Xie AJ, Shen YH, et al. Fabrication of composite photocatalyst g-C_3N_4-ZnO and enhancement of photocatalytic activity under visible light. Dalton Transactions. 2012;**41**:6756-6763. DOI: 10.1039/c2dt12474b

[191] Wang Y, Wang Z, Muhammad S, He J. Graphite-like C_3N_4 hybridized $ZnWO_4$ nanorods: Synthesis and its enhanced photocatalysis in visible light. CrystEngComm. 2012;**14**:5065-5070. DOI: 10.1039/c2ce25517k

[192] Yuan D, Huang L, Li Y, Xu Y, Xu H, Huang S, et al. Synthesis and photocatalytic activity of g-C_3N_4/BiOI/BiOBr ternary composites. RSC Advances. 2016;**6**:41204-41213. DOI: 10.1039/C6RA05565F

[193] Dong Z, Wu Y, Thirugnanam N, Li G. Applied surface science double Z-scheme ZnO/ZnS/g-C_3N_4 ternary structure for efficient photocatalytic H_2 production. Applied Surface Science. 2018;**430**:293-300. DOI: 10.1016/j.apsusc.2017.07.186

[194] Gao Y, Lin J, Zhang Q, Yu H, Ding F, Xu B, et al. Facile synthesis of heterostructured YVO4/g-C_3N_4/Ag photocatalysts with enhanced visible-light photocatalytic performance. Applied Catalysis B: Environmental. 2018;**224**:586-593. DOI: 10.1016/j.apcatb.2017.11.003

[195] Zhong Y, Yuan J, Wen J, Li X, Xu Y, Liu W, et al. Earth-abundant NiS co-catalyst modified metal-free mpg-C_3N_4/CNT nanocomposites for highly efficient visible-light photocatalytic H_2 evolution. Dalton Transactions. 2015; **44**:18260-18269. DOI: 10.1039/c5dt02693h

[196] Yan J, Song Z, Wang X, Xu Y, Pu W, Xu H, et al. Enhanced photocatalytic activity of ternary Ag_3PO_4/GO/g-C_3N_4 photocatalysts for rhodamine B degradation under visible light radiation. Applied Surface Science. 2019;**466**:70-77. DOI: 10.1016/j.apsusc.2018.09.234

[197] Zhou Y, Li J, Liu C, Huo P, Wang H. Construction of 3D porous g-C_3N_4/AgBr/rGO composite for excellent visible light photocatalytic activity. Applied Surface Science. 2018;**458**:586-596. DOI: 10.1016/j.apsusc.2018.07.121

[198] Jiang D, Yu H, Yu H. Modified g-C_3N_4/TiO_2 nanosheets/ZnO ternary facet coupled heterojunction for photocatalytic degradation of p-toluenesulfonic acid (p-TSA) under visible light. Physica E: Low-dimensional Systems and Nanostructures. 2017;**85**:1-6. DOI: 10.1016/j.physe.2016.08.014

[199] Yuan Y, Huang G, Hu W, Xiong D, Zhou B, Chang S. Construction of g-C_3N_4/CeO_2/ZnO ternary photocatalysts with enhanced photocatalytic performance. Journal of Physics and Chemistry of Solids. 2017; **106**:1-9. DOI: 10.1016/j.jpcs.2017.02.015

[200] Deng Y, Wang J, Liu Y, Wang J, Feng H, Tang L, et al. Enhanced photocatalytic activity of ternary Ag/g-C_3N_4/$NaTaO_3$ photocatalysts under wide spectrum light radiation: The high potential band protection mechanism. Applied Catalysis B: Environmental. 2018;**230**:102-114. DOI: 10.1016/j.apcatb.2018.02.031

Chapter 3

TiO_2 Nanoparticles Supported on Hierarchical Meso/Macroporous SiO_2 Spheres for Photocatalytic Applications

Keyla M. Fuentes, Margarita Sánchez-Dominguez and Sara A. Bilmes

Abstract

Supporting a photocatalyst, such as titania nanoparticles (TiO_2 NPs), is a good strategy to improve its performance since it can facilitate the photocatalyst recovery from the aqueous media and provides a high surface area for pollutant adsorption. Among the several advanced functional materials used as TiO_2 NP support, the hierarchical meso/macroporous SiO_2 spheres not only show the advantages associated to its chemical nature but also the dendritic fibrous structure provides a porous network that offers many benefits to be exploited in optical and catalytic devices. In this chapter, different synthetic approaches to design hierarchical meso/macroporous silica and the strategies to support TiO_2 NPs regarding the photocatalytic performance of these materials are shown.

Keywords: hierarchical SiO_2 spheres, meso/macroporous network, dendritic silica, supported TiO_2 nanoparticles, TiO_2 photocatalysis

1. Introduction

The practical applications of semiconductor photocatalysis using TiO_2 nanoparticles (TiO_2 NPs) are still limited by the catalyst recovery from the aqueous media. Supporting TiO_2 NPs on suitable materials is an advantageous strategy to overcome this issue. The effectiveness of this process requires that the supporting material has mechanical and thermal strength, high surface area for the loading of NPs, and surface sites for covalent binding to avoid NPs detachment. Nevertheless, the support must not modify the electronic structure of TiO_2 nor reduce the density of the surface defects that promotes the adsorption of reactive molecules [1–3]. Silica-based materials are attractive supports for TiO_2 due to several reasons: (i) it enhances the thermal stability of the photocatalytically active anatase phase, (ii) it avoids the particles agglomeration [4], (iii) the establishment of an Si-O-Ti interaction prevents the NPs detachment, and (iv) SiO_2 is transparent in UV-vis region and it does not interfere with the generation of electron-hole pairs in TiO_2 [5]. Additionally, SiO_2@TiO_2 core@shell structures enhance the absorption of photons due to the contrast between the refractive index producing a light-harvesting effect, which favors the photoactivity [6].

The optimal TiO_2 photocatalysts are 5–50 nm NPs in the anatase phase [7]. High loadings of NPs within this size range could be challenging for supports with low surface areas. Mesoporous silica with high surface area (500–1000 m^2/g), such as MCM-41, MCM-48, SBA-15, and KIT-6 has been widely used to deposit TiO_2 NPs [8–11]. However, the pore diameters in these materials usually range from 2 to 15 nm [12]; hence, the crystallization of NPs with sizes comparable to the pores entrances will cause, inevitably, the clogging of the pores and a significant decrease in the surface area [13].

Hierarchically porous materials have the advantage that micro- and mesopores provide the size and shape selectivity for guest molecules, enhancing the host-guest interactions, whereas macropores favor the diffusion and accessibility of guest molecules [14]. The need for tailored porous materials for different applications has motivated the design of hierarchical meso/macroporous silica by different synthetic approaches [15]; most of them are based on the sol-gel chemistry at liquid interfaces using arrangements of amphiphilic molecules as templates [16–18]. Interesting examples of these materials are the so-called dendritic fibrous nanosilicas (DFNS) [19–21], such as KCC-1 spheres (KAUST Catalysis Center-1) that are potentially useful for solving problems in catalysis, photocatalysis, DNA adsorption, CO_2 capture, optical devices, and drug delivery [22].

For photocatalytic applications, the location of the TiO_2 NPs plays an important role in the photoactivity, since NPs should be accessible for both, photons and reactive molecules. **Figure 1** illustrates the different locations of the TiO_2 NPs deposited on the fibrous silica KCC-1 in comparison with conventional mesoporous silica such as MCM-41 or SBA-15. In case of NPs deposited on KCC-1, the broad size of pore entrances facilitates the accessibility to the particles located closer to the center of the sphere.

In this chapter, we describe the synthesis pathways and the strategies for turning fibrous silica as support for photocatalytic TiO_2 NPs for applications in dye-sensitized solar cells, water splitting, and photodegradation of dyes. The advantages of using these materials as support for TiO_2 in relation to photocatalytic processes are addressed by comparing the photocatalytic activity of TiO_2 NPs deposited by different routes, as well as with TiO_2 NPs on mesoporous materials. Since the structure of these materials in the nanometer scale presents many analogies with forms of

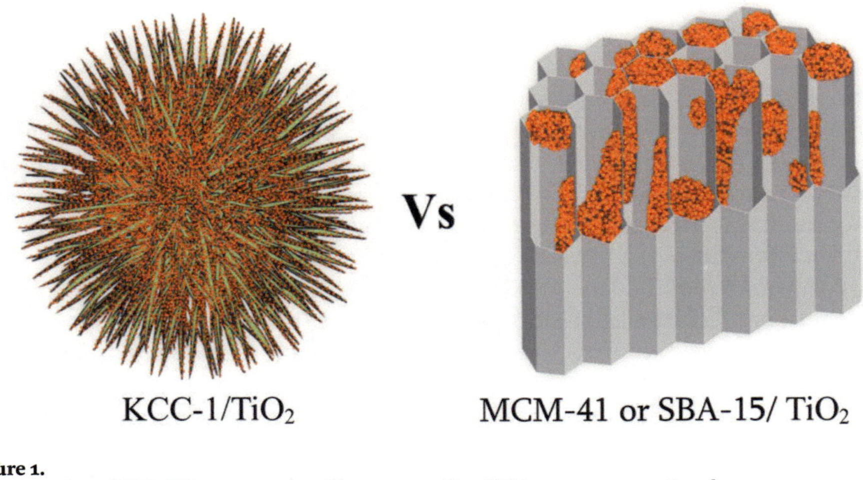

Figure 1.
Representation of TiO_2 NPs supported on fibrous nanosilica KCC-1 versus conventional mesoporous support (e.g., MCM-41 or SBA-15). The orange dots symbolize the TiO_2 NPs. Reprinted with permission from Reference [23]. Copyright 2016 American Chemical Society.

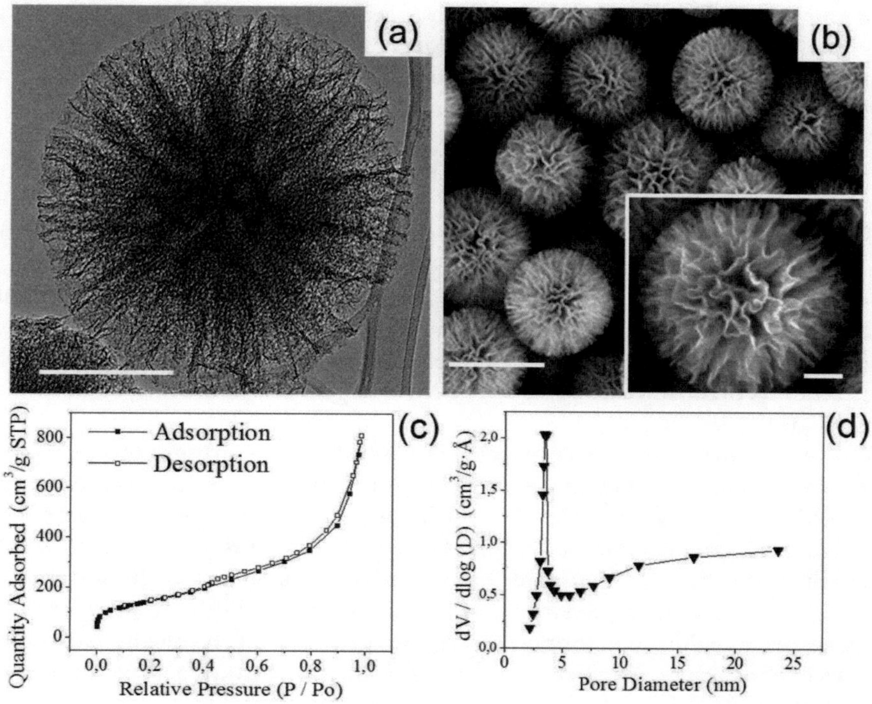

Figure 2.
TEM (a) and SEM (b) images for a representative hierarchical meso/macroporous SiO$_2$ spheres obtained through the synthesis by type III Winsor emulsion (see below). For the TEM image, the scale bar is 200 and 500 nm for the SEM image (scale bar of 100 nm for the inset). (c) N$_2$ adsorption—desorption isotherm and (d) pores size distribution (calculated by the BJH model) for the particles shown in (a and b). Unpublished results of the authors.

the micro- and macroscopic world: dendrites, carnations, wrinkles, fibers, in what follows, we will use indistinguishably the terms fibrous, dendritic, or wrinkled to refer to structures like those in **Figure 2**.

2. Synthesis of the hierarchical meso/macroporous SiO$_2$ spheres

Most of the novel methods for building nanostructures with tunable morphology and pore size are based on the application of an oil—water interface using amphiphilic molecules as template, which allows the preparation of materials with well-defined structures in a confined reaction media [24, 25]. By changing the oil phase, surfactant, or reaction parameters, several morphologies can be obtained, such as rod-like [26], vesicle-like [27], cage-like [28], or fibrous structures [29].

The morphological and textural characteristics of a fibrous structure are shown in **Figure 2**. Transmission electron microscopy (TEM) image reveals a darker core surrounded by a meso/macroporous shell (**Figure 2a**). Unlike the narrow pore-size distribution of common mesoporous silica materials (e.g., MCM-41 and SBA-15), these particles have radially oriented pores whose diameter increases from the center of the sphere. The scanning electron microscopy (SEM) images show a narrow distribution of particle size, and the surface of each sphere presents cavities with a broad distribution of sizes limited by folded walls (**Figure 2b**). The associated nitrogen adsorption—desorption isotherm shown in **Figure 2c** could be classified as type IV isotherm with H$_3$ hysteresis loop. Materials with H$_3$ hysteresis do not show any limiting adsorption at high *P*/*P*o associated with nonrigid aggregates of plate-like particles and have slit-shaped pores [30]. These particles have a complex porous structure (**Figure 2d**) with pores of 3.5 nm, resulting from the surfactant assemblies that define the pore structure of the walls [31], as derived from BJH analysis of adsorption—desorption isotherms. Additionally, there is a broad distribution of larger pores related to the lamellar structure unfolding toward the particle surface.

This type of morphology with radially oriented channels can be obtained in a ternary confined reaction media composed by a surfactant and a co-surfactant, a hydrocarbon, and water. In case of particles presented in **Figure 2**, the reaction media is composed by cetyltrimethylammonium bromide (CTAB) and 1-butanol, cyclohexane, and water. Usually, the silicon sources are alkoxides, such as tetraethyl or tetramethyl orthosilicates (TEOS, TMOS) that give rise to the SiO_2 structure through sol-gel reactions.

The water-surfactant (+co-surfactant)-oil system may present one or more phases depending on the chemical composition. For instance, a thermodynamically stable microemulsion phase or multiphase systems containing a microemulsion layer can be obtained [32]. To follow the description of the formation mechanism, it is necessary to introduce some concepts related to the system.

Micelle: structures formed by self-assembly of amphiphiles in aqueous solutions giving separate regions with opposite affinities toward a given solvent. Normally, micelles have characteristic size within the 2–50 nm range and form spontaneously under certain conditions of concentration and temperature [33].

Reverse micelle: spheroidal aggregates formed by amphiphiles in organic solvents. They can be formed both in the presence and in the absence of water. In a medium free of water, the aggregates are small and polydisperse [34].

Microemulsion: system composed of water, oil, and an amphiphile, which is a single optically, isotropic, and thermodynamically stable liquid solution [35]. The microemulsions can exist as oil-swollen direct micelles dispersed in water (O/W microemulsions), water-swollen reverse micelles dispersed in oil (W/O microemulsions), or with both aqueous and oily continuous domains as interconnected sponge-like channels (bicontinuous microemulsions) [36].

Winsor emulsion systems: the Winsor classification of ternary systems, i.e., types I–IV, can be made according to the numbers and type of phases in the equilibrium state. Except for the single-phase microemulsion (type IV), the structure of the microemulsion is determined through the separation of the solvent from the dispersed phase of the microemulsion layer. Type I: at equilibrium, superfluous oil separates from the oil-in-water microemulsion. Type II: superfluous water separated from the microemulsion layer. Type III: both superfluous oil and water separated from the water-in-oil microemulsion [32].

Mesophase: a phase occurring over a defined range of temperature, pressure, or composition within the mesomorphic state. This is a state of matter in which the degree of molecular order is intermediate between the three-dimensional, long-range order found in solid crystals and the order found in isotropic liquids, gases, and amorphous solids [37].

Sol-gel reactions: these are kinetically controlled hydrolysis and polycondensation reactions of a metallic precursor (e.g., inorganic or alkoxide), which give rise to a metal oxide structure. The precursor, building blocks, evolves in several steps from dispersions of colloidal particles in a liquid (Sol) to interconnected polymeric chains forming a porous rigid network (gel) [38]. The most generally accepted model to describe silica polymerization in aqueous media is nucleation and growth [39].

2.1 Proposed mechanism for the formation of the hierarchical porous SiO_2 using oil-in-water emulsions

The general mechanism for the formation of hierarchical porous silica using oil-in-water emulsion made from a type III Winsor system can be described in the following steps, also schematized in **Figure 3** [20]:

1. The silicon source dissolved in the oil layer meets the water layer—an aqueous solution of a base—at the emulsion interface; the sol-gel reactions take place at this stage. In the basic solution, the reaction mixture contains ionized silicate monomers and oligomers negatively charged. As the silica condensation proceeds, the amount of partially condensed silicates decreases and that of fully condensed silicates increases.

2. Silicate oligomers interact with the surfactant heads giving rise to a change in the curvature of the interface.

3. As the curvature increases, a closed structure such as spherical or cylindrical shapes can be formed. These surfactant-silicate systems with closed structure lead to the formation of micelles or micellar emulsions inside the channels of the bicontinuous microemulsion.

4. The aggregation of these micelles leads to the formation of a repetitive mesophase.

However, in this mechanism, the growth of the particle is still unclear. It can be thought that the formed mesophase may act as a seeding site were the wrinkles grow through the water channels. As particles start growing, the structure of the microemulsion is rearranged, and according to Moon and Lee [20], the walls of wrinkles are formed in the water layers and the valleys are consequences of the oil phase in the microemulsion.

An alternative mechanism was proposed by Febriyanti et al. [40], considering the optimal conditions for the synthesis of KCC-1 [29]. They suggest the formation of reverse micelles in which the sol-gel reactions take place at higher rate in the center of the droplet. This mechanism is schematized in **Figure 4**. Although this mechanism explains the TEM images in which the density of the particles seems to decrease from the center to the surface of the sphere, experiments using real-time small angle X-ray scattering and 3D tomography demonstrate that the fibers grow uniformly from the

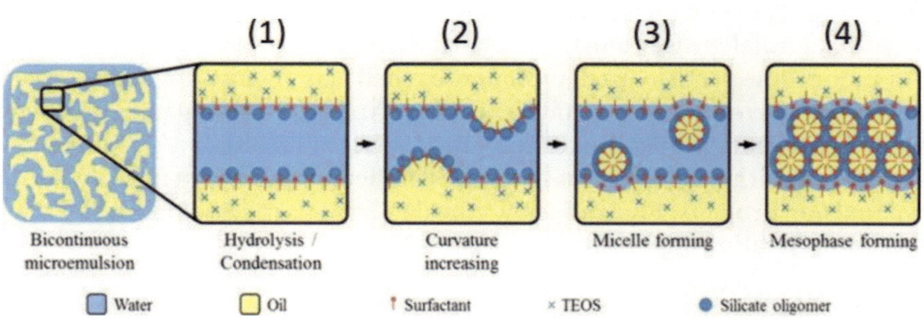

Figure 3.
Schematic illustration of the mesophase-forming mechanism from the microemulsion interface. Reprinted with permission from Reference [20]. Copyright 2012 American Chemical Society.

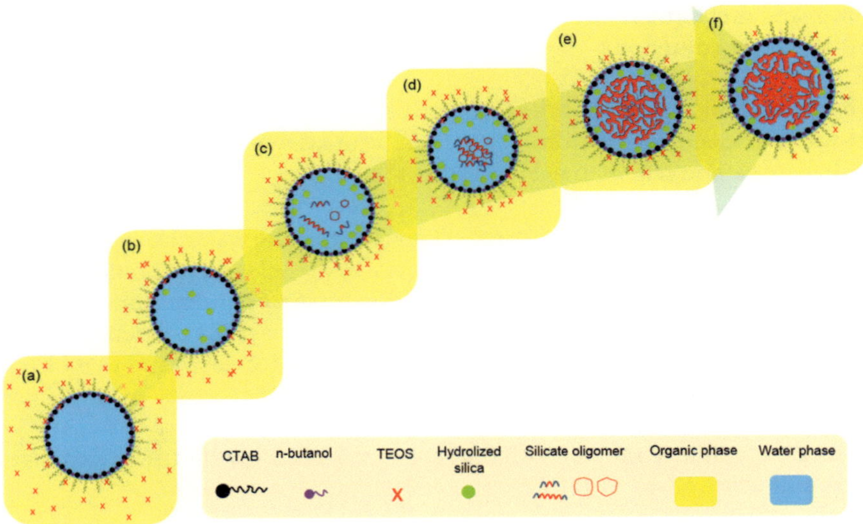

Figure 4.
Schematic representation of the KCC-1 synthesis in the reverse emulsion system proposed by Febriyanti et al. Reprinted with permission from Reference [36]. Copyright © 2016, American Chemical Society.

center of the spheres along the free radial directions to form fibrous spherical silica particles. Based on these results, Maity et al. [41] described the microemulsion-droplet-coalescence mechanism.

2.2 Tuning the morphology and pore size distribution of the hierarchical meso/macroporous silica spheres

The possibility to obtain tailor-made supports for specific guest molecules and tunable pore volume for loading control is the most attractive feature of the three-phase synthesis approach. In case of fibrous nanosilica KKC-1, the surface area may vary from 450 to 1244 m^2/g, and the tunable pore sizes (3.7–25 nm) allow to increase the pore volume up to 2.18 cm^3/g. Likewise, the particle size can be controlled from 40 to 1120 nm. [42]. Parameters controlling particle size, pore volume, and surface area are discussed below.

2.2.1 Oil phase

Shen et al. developed a stratification approach to obtain a three-generation pore size by changing the oil phase in a biphasic system [43]. They used 1-octadecene, decahydronaphthalene (decalin), and cyclohexane to pores of sizes 2.8, 5.5, and 7.0 nm in the same particle. Each mesoporous shell was obtained in a synthesis step using the corresponding oil phase. This confirms that the organic solvent not only provides a storage medium for the silicon source (TEOS), but also interacts and assembles with the surfactant molecules in the interface to form oil-in-water microemulsion droplets. **Figure 5** shows the dendritic particles obtained by the biphasic stratification approach.

2.2.2 Surfactant

Depending on the molar ratio of surfactant to silicon, the surfactant participates in the synthesis with different functional roles: controlling the particle diameter, the particle dispersion in the medium, and the degree of mesophase formation.

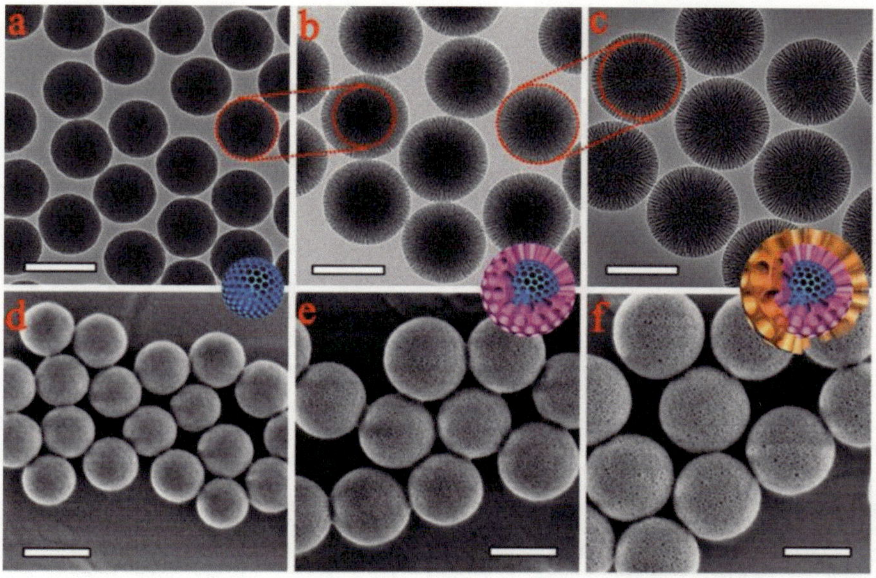

Figure 5.
TEM (a-c) and SEM (d-f) images of the extracted 3D-dendritic MSNSs with one (a and d), two (b and e), and three generations (c and f) prepared via the biphase stratification approach. All scale bars in TEM and SEM images are 200 nm.

As the surfactants favor the dispersion of the alkoxysilanes in water, the hydrolysis reactions are favored in comparison with the condensation reactions leading to an increased nucleation rate. The net effect is that increasing the surf/Si ratio, the particle diameter decreases. Moreover, at a high surf/Si ratio, free micelles may adsorb in the primary particles, allowing the dispersion of the particles due to electrostatic and/or steric repulsion. In contrast, at low surf/Si ratios, the surfactants mainly act as porogen [44]. For the KCC-1 synthesis using benzyldimethylhexadecyl ammonium chloride (16- BAC) instead of CTAB, the fibrous morphology is lost due to the differences in the packing parameters, which in turn depend on the length of the nonpolar chain and the effective area of the polar head [29].

2.2.3 Co-surfactant

By increasing the alkyl chain length of the co-surfactant (linear C_3-C_{20} alcohols), the distance between wrinkle walls increases, as shown in **Figure 6**. As the affinity between oil and surfactant defines the valleys of the wrinkles, an increase in the hydrophobicity of the aqueous phase by enlarging the co-surfactant alkyl-chain broadens the distance between the wrinkle walls. It can be considered that this distance is directly related to the density of fibers by changing this parameter, the pore size distribution in the region related to the lamellar structure and the pore volume also changes, but the surface area remains unaltered [20].

2.2.4 Composition of the water phase

The type and concentration of the base catalyst dissolved in the aqueous phase affects the diameter of the spheres, as it controls the hydrolysis rate of the silicon source. Due to their buffer capacity, weak bases are preferred. For instance, basic amino acids, such as arginine or lysine, keep the reaction under weakly basic conditions leading to spherical particles with dendritic pores [18, 45]. The same particles

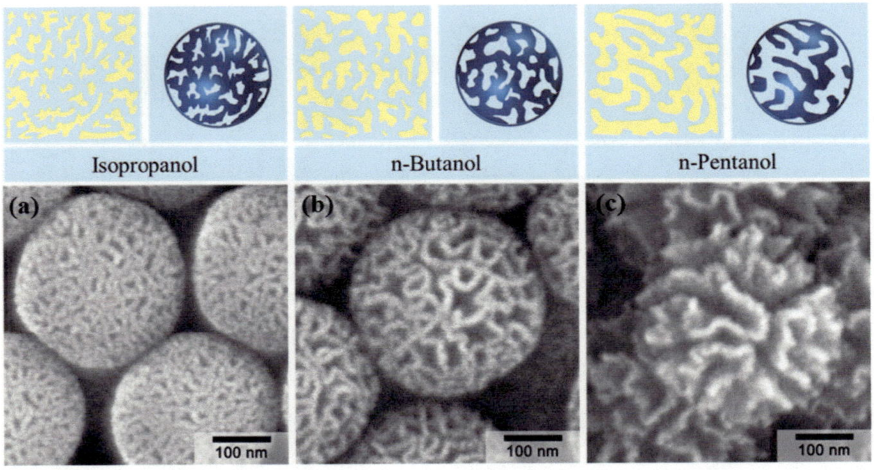

Figure 6.
Schematic illustrations of microemulsion phases of reaction mixtures with different co-solvents (6.0 mmol) and corresponding SEM images of the resulting wrinkled SiO_2 spheres: (a) isopropanol, (b) n-butanol, and (c) n-pentanol. Reprinted with permission from Reference [20]. Copyright 2012 American Chemical Society.

can be obtained with urea, whose thermal decomposition above 70°C maintains the solution pH between 7 and 9.5, depending on the urea concentration [46]. Nevertheless, the increase in urea concentration rises the nucleation rate leading to smaller particles.

2.2.5 Stirring

Jin et al. [47] found that keeping constant the emulsion composition to obtain rod-like structures at low stirring speeds (i.e., <300 rpm) produces twisted-ribbon or twisted-rod-like structures with changes in the length and the diameter of the rods. On the other side, high stirring speeds (i.e., 1200 rpm) yield spherical or irregular morphologies. The stirring speed has two main effects in the synthesis: (i) it alters the self-assembly of the template and (ii) it affects the diffusion of the reagents to the interface [17]. In the synthesis of wrinkled SiO_2, high stirring speeds (around 900 rpm) are preferred to obtain a spherical morphology.

2.2.6 Reaction time

Hydrolysis and condensation rates for silica at pH 7–9 are slow processes occurring simultaneously with changes in surfactant assemblies. The synthesis of dendritic particles in the conditions giving the product as depicted in **Figure 2** requires around 20 h to be completed. The particle evolution during these conditions monitored by SEM is shown in **Figure 7**. After 4 h, slightly structured particles ($\leq$100 nm) are formed. As the reaction proceeds, lamellae are defined and particles grow, reaching an average diameter of 300 nm with high size dispersion after 6 h. Once lamellae are formed, the effect of reaction time is a slow increase of particle size up to 500 nm in 24 h, less polydispersion, and well-defined planes of wrinkles with deep valleys.

Several processes converge to achieve these corrugated particles. On the one hand, the silicon precursor condenses on the already formed silica units. On the other hand, there is a simultaneous silicate dissolution and re-precipitation (Oswald ripening) that generates hydrophilic silica surfaces. This process induces the formation of hydrophobic regions where the organic phase is accumulated giving

Figure 7.
Evolution of the growth in time of the particles shown in ***Figure 2*** *monitored by SEM. Scale bars are 200 nm. Unpublished result from authors.*

rise to the deep valleys observed for 24 h. Although data on the kinetics of dendritic particles formation are scarce, it is known that the reaction may be accelerated by the assistance of microwave radiation, as proposed for the synthesis of the KCC-1. Under these conditions, particles are formed in few minutes but for longer reaction times (>60 min); the lamellae collapse producing denser particles [42]. This experiment reinforces the idea that the necessary condition for having a dendritic structure is the formation of hydrophobic "pockets," a process that is inhibited when microwave radiation is applied.

In summary, the main advantage of the synthesis using oil-in-water Winsor III emulsion is the production of dendritic particles that can be tailored by the synthesis parameters for a specific need. The adjustment of pore sizes is important for trapping biomolecules or nanoparticles and for diverse catalytic applications that require high surface area and pore volume. Moreover, for environmental applications, the particle size is a key factor for catalyst recovery. The surfactant to Si ratio and the concentration of the base catalyst are important parameters for controlling the particle diameter, whereas the microemulsion phase behavior defines the distance between the wrinkle walls, i.e., the macroporosity. For a given surfactant, the microemulsion structure is affected by the organic solvent and the co-surfactant chain length, which will tailor the size of cavities and fiber density. The reaction time is an empiric variable for each reaction set-up, which needs to be optimized to get the required product.

3. Strategies for supporting TiO_2 NPs on hierarchical meso/macroporous silica spheres and photocatalytic performance

Catalysts based on Ti-containing mesoporous silica, e.g., MCM-41. SBA-15 has been developed from different synthetic pathways for applications such as selective oxidation of bulky organic compounds and photoreduction of greenhouse gases [48, 49].

Typically, two synthesis strategies could be applied depending on the stage of incorporation of the TiO_2 precursor: (i) *one-pot*: added directly to the SiO_2 reaction mixture (co-condensation route) or (ii) *postsynthesis*: added after the consolidation of the SiO_2 network by deposition, impregnation, atomic layer deposition, among others. In the one-pot approach, the co-condensation of titanium precursor with the silicon source faces with the problem that Ti alkoxides are much more reactive than $Si(OR)_4$ [50]; thus, it is necessary to control the Ti precursor hydrolysis in order to have an appropriate distribution of Ti species that lead to the formation of anatase nanocrystals, a necessary component for efficient photocatalysis [51].

In this section, we focus on different postsynthesis pathways leading to photocatalytic systems for energy conversion, H_2 production, or photodegradation of dyes. The consolidated SiO_2 porous network must ensure high loadings of homogeneously distributed TiO_2 NPs attached by covalent interactions, while the deposition method determines the location and structure of the nanoparticles, e.g., particle size and crystallinity. All these features affect the photocatalytic performance of the semiconductor.

TiO_2 NPs deposited on KCC-1 by atomic layer deposition (ALD) alternating titanium tetra-butoxide and hydrogen peroxide produces anatase NPs of 3–12 nm highly dispersed on the fibers (**Figure 8a**) whose diameter increases with the number of ALD cycles [21]. The anatase phase was obtained after the ALD cycles by calcination at 300–700°C. This material exhibits high photocatalytic activity and turnover frequency (TOF) towards the Rhodamine B (Rh-B) photodegradation under UV light, even better than the one obtained using the same ALD procedure with MCM-41 and SBA-15 substrates (**Figure 8b** and **c**). This higher photoactivity may be a consequence of the location and distribution of TiO_2 NPs on the support allowing an optimal dye-TiO_2 interaction for a hole transfer or improved OH• attack

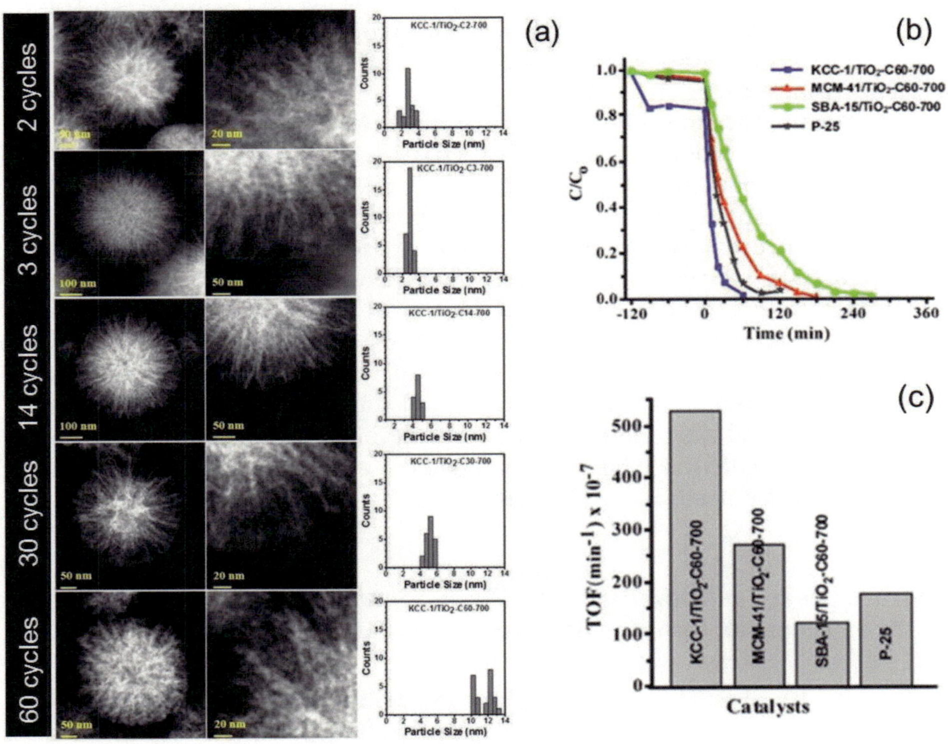

Figure 8.
(a) Scanning transmission electron microscopic images (STEM) of KCC-1/TiO_2 after several ALD cycles and heat treatment at 700°C, the corresponding histogram for the average particle size is showed next to each image. (b) Photocatalytic degradation of rhodamine B using KKC-1/TiO_2 compared to equivalent catalyst using common mesoporous silica and a commercial Degussa P25. (c) Turnover frequency (TOF) for the reaction: it was estimated as the moles of dye degraded per mole of TiO_2 per minute. Adapted from Reference [23]. Copyright 2016 American Chemical Society.

to the dye. The open structure of the fibrous silica admits higher TiO_2 loading accessible to the dye, while the ordered array of mesoporous channels in MCM-41 and SBA-15 is easily clogged by TiO_2 NPs as schematized in **Figure 1**. The TiO_2 loading on SiO_2 is determined by the surface silanol/siloxane ratio which is independent on synthesis method [52]. Thus, even all substrates exhibit nearly the same TiO_2 loading in weight (**Figure 9**), the more open structure of KCC-1 allow better accessibility to Ti precursors as ALD cycles increase.

Enhanced photocatalytic activity can also be explained by higher photon harvesting in the fibrous structure of the KCC-1 [42], as these structures enable incident light to pass through the cavities and reduce optical loss due to multiple light scattering and reflection [53]. The efficient photon harvesting of dendritic particles was also exploited for the design of dye-sensitized solar cells (DSCs) [48]. By modifying the interwrinkle distances in the silica spheres with the co-surfactant's length of the carbon chain, the fiber density changes from narrow (NWSNs) to wide (WWSNs) wrinkled silica. In this case, TiO_2 NPs were deposited by dispersing the SiO_2 spheres in a mixture of acetonitrile- ethanol (1:1) using ammonia solution to conduct the hydrolysis/condensation reactions of titanium ter-butoxide. Using this approach, 4 nm anatase crystallites were obtained after calcination at 500°C under air atmosphere. The dependence of photovoltaic parameters with the fiber density is interpreted as a consequence of the effective refractive index resulting from the combination of two materials with different optical properties, supporting the idea of improved photon harvesting promoted by the fibrous nature of the SiO_2 spheres.

In the latter cases, the crystallization of TiO_2 was conducted by calcination at temperatures higher than 300°C. Despite the fact that porous silica spheres have good thermal stability, the calcination presents several drawbacks for the photocatalytic process: (i) it reduces the surface hydroxyl groups and surface area [54] and (ii) it induces the anatase to rutile transformation leading to a decrease in photoactivity [55]. Although anatase-rutile phase transition can be managed with the calcination temperature and by the introduction of defects, hydrothermal synthesis of TiO_2 NPs is a more convenient method for producing anatase nanocrystals on silica substrate [6].

On the hydrothermal route, the solvent has a strong impact on the TiO_2 NP size and distribution [56]. Under soft conditions (180°C, 20 h) in ethanol-water mixtures, the TiO_2 NPs are homogeneously located in the outermost zone of KCC-1 particles when using pure water. This causes an increment in the specific surface area and pore volume; whereas, by using absolute ethanol, the particles are dispersed in the fibers (see TEM images and particle size distribution in **Figure 10a**). Regarding

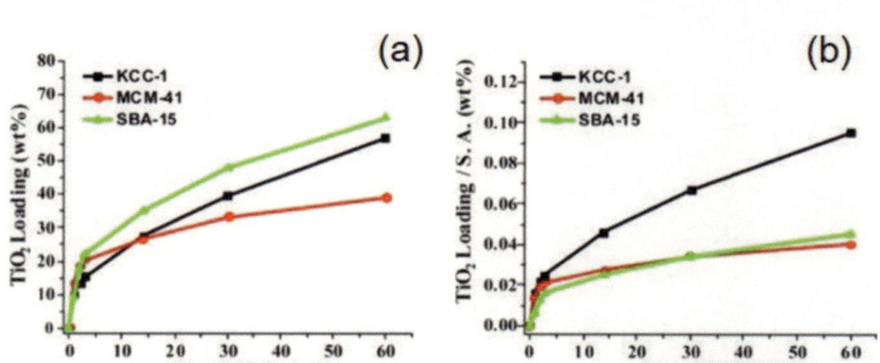

Figure 9.
TiO_2 loading on KCC-1, MCM-41, and SBA-15 deposited by ALD. The loading is expressed as wt% (a) and wt% per unit surface area (b). Reprinted from Reference [23]. Copyright 2016 American Chemical Society.

the TiO_2 NPs sizes, it goes from 8 nm to 17.5 nm by increasing the ethanol/water ratio from 0/100 to 28.5/1.5. Then, it decreases to 5 nm in a 100/0 ratio; under this condition, a narrower size distribution is obtained. As far as photocatalytic activity is concerned, particles prepared with absolute ethanol showed the highest activity for water splitting under UV light (**Figure 10b**). This result points out that the size and location of the TiO_2 NPs plays a significant role in the photocatalytic performance. The TiO_2 NP distribution leads to different morphologies with a characteristic effective refractive index, which defines the interaction with the incident photons. Thus, by controlling the accessibility of the TiO_2 NPs to the porous system, it is possible to design suitable morphologies for improving photocatalysis.

Another alternative to control the TiO_2 NP location is by taking advantage of the reagents used during the synthesis of the support. As shown in the previous section, the synthesis of wrinkled SiO_2 spheres (see **Figure 2**) requires high amounts of CTAB acting as a pore template. Thus, it is possible to imagine that before removing the template, the pores are "filled" with CTAB molecules blocking the transport of Ti precursors. On this basis, our group developed a practical approach to handle the TiO_2 NP distribution on wrinkled SiO_2 spheres (460 nm in diameter, B.E.T. area = 550 m^2/g) produced by hydrothermal synthesis using Ti isopropoxide in isopropanol. With this hydrothermal synthesis, TiO_2 NPs of 10 nm with 6-nm anatase crystallites are obtained. The loading and location of the NPs significantly differ depending on the presence of CTAB during the hydrothermal treatment. **Figure 11a** summarizes the experimental approach for controlling the TiO_2 NP location and their corresponding TEM and SEM images (**Figure 11b**). In **Figure 11**, *w*-SiO_2@TiO_2 refers to NPs deposited after the removal of CATB and w-SiO_2/CTAB@TiO_2 to NPs deposited in the spheres containing CTAB. By eliminating CTAB prior to TiO_2 deposition, NPs are highly dispersed in a hybrid SiO_2-TiO_2 shell, resulting in a greater decrease in surface area than when TiO_2 NP growth is performed in the presence of CTAB. This morphology resembles that

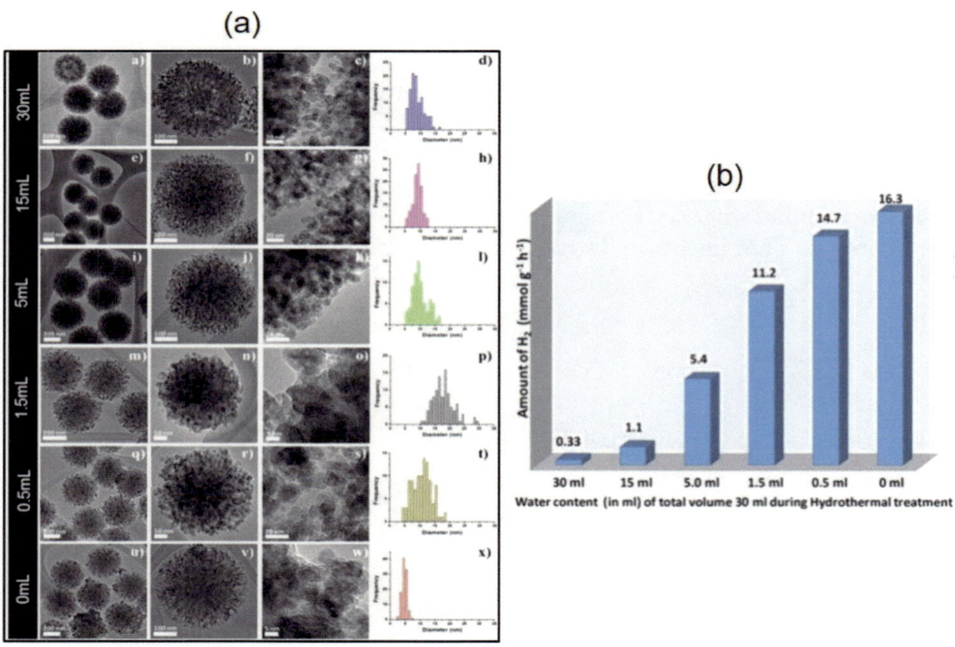

Figure 10.
(a) TEM images in three magnifications and corresponding histograms for the particle size distribution after hydrothermal treatment at 180°C obtained in varying water content (inset: a-x). (b) H_2 evolution using the samples showed in the SEM images [56].

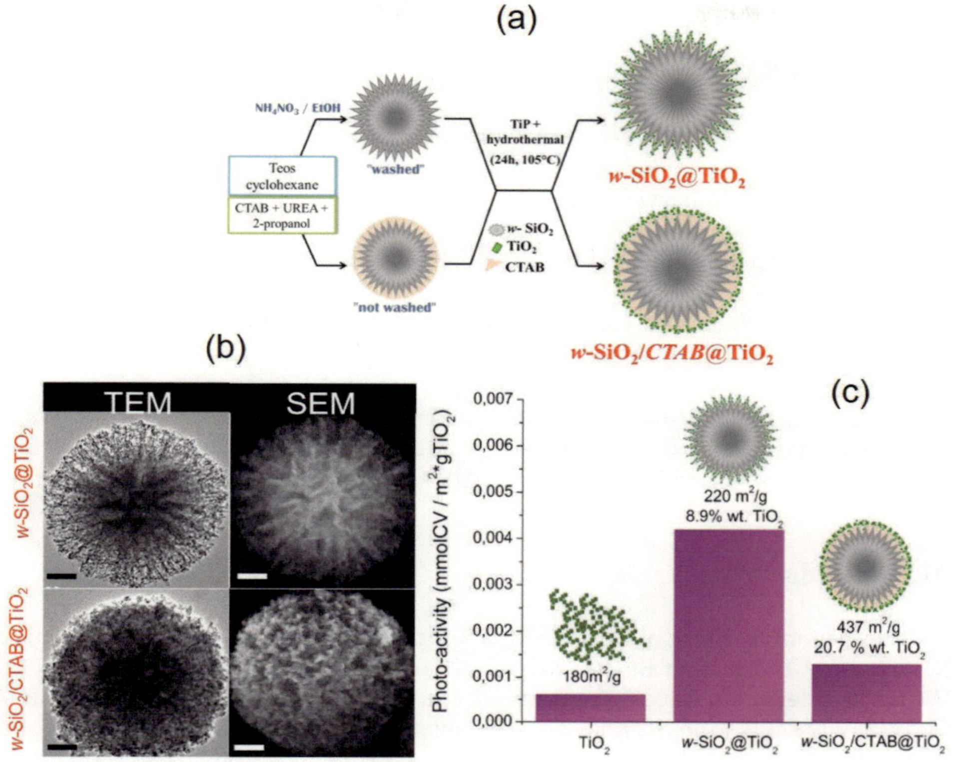

Figure 11.
(a) Experimental scheme for the synthesis of wrinkled-SiO_2@TiO_2 particles, (b) corresponding TEM and SEM images, the scale bar are 100 nm, and (c) photoactivity of the wrinkled-SiO_2@TiO_2 particles in comparison with the unsupported TiO_2 NPs. Unpublished results of the authors.

obtained by atomic layer deposition (ALD) [21], as shown in **Figure 8a**. On the other side, when the hydrothermal process is carried out in the presence of CTAB, the NPs are arranged as a mesoporous TiO_2 layer in the same way as the particles obtained by hydrothermal crystallization described in **Figure 10a** [51]. Therefore, this synthetic approach provides different morphologies simply by taking advantage of CTAB molecules from the synthesis of SiO_2 particles. The photoactivity of these particles was tested in the degradation of crystal violet (CV). As seen in **Figure 11c**, CV photodegradation occurs at higher rate when using SiO_2-TiO_2 hybrid layer particles (w-SiO_2@TiO_2) whose morphology allows better photon harvesting. In case of w-SiO_2@CTAB TiO_2 particles, the presence of CTAB increases the adsorption of the dye, thus providing a basis to functionalize the SiO_2 particles for a more specific interaction with the targets. Finally, the dimensions of the support are adequate for a good recovery of the reaction medium providing an efficient system for environmental applications.

4. Summary and outlook

Meso/macroporous silica spheres are remarkable supports for TiO_2 NPs as compared to conventional silica mesoporous materials, such as MCM-41 and SBA-15. The main advantage of this morphology is the accessibility of the molecules to the silica surface provided by the broad pore size distribution. For photocatalytic applications, the location of TiO_2 nanoparticles is a critical parameter to optimize the photoactivity. Supported TiO_2 NPs in these materials exhibit:

i. better accessibility to the active sites ensuring high TiO_2 loadings;

ii. high surface area, which favors higher adsorption of the pollutants;

iii. enhanced photocatalytic performance for several target molecules;

iv. improved light harvesting; and

v. facile recovery from the reaction media compared to unsupported TiO_2 NPs.

The easy tailoring of the surface properties provides photocatalysts for specific applications that can be rationalized by considering the location of TiO_2 NPs on the porous support, which in turn can be managed by the synthesis variables. Understanding the internal nature of the hierarchical supports of the photoactive TiO_2 allows the development of value in photocatalysis for energy and environmental applications—a contribution of material's scientists to a better world.

Acknowledgements

The authors are grateful with CONICET for the postdoctoral scholarship of K.M.F., and University of Buenos Aires for funding, UBACyT 20020170200298BA. We are in indebted to Nanoandes Network and ERANET-LAC FP7 project Recola for the continuous support. Also, we acknowledge Stefanie Kodjikian (Institut NÉEl, Grenoble) and Maria Claudia Marchi (CMA-CONICET) for TEM and SEM images, respectively.

Conflict of interest

There are no conflict of interest to declare.

Author details

Keyla M. Fuentes[1,2]*, Margarita Sánchez-Dominguez[2] and Sara A. Bilmes[1]

1 Instituto de Química Física de los Materiales, Medio Ambiente y Energía-(INQUIMAE-CONICET), Departamento de Química Inorgánica, Analítica y Química Física (DQIAQF), Facultad de Ciencias Exactas y Naturales, Universidad de Buenos Aires, Ciudad Universitaria (C1428EHA), Ciudad Autónoma de, Buenos Aires, Argentina

2 Centro de Investigación en Materiales Avanzados, S.C. (CIMAV), Apodaca, Nuevo León, México

*Address all correspondence to: keylafuentesflores@gmail.com

References

[1] Miar Alipour S, Friedmann D, Scott J, Amal R. TiO_2/porous adsorbents: Recent advances and novel applications. Journal of Hazardous Materials. 2018;**341**:404-423. DOI: 10.1016/j.jhazmat.2017.07.070

[2] Schneider J, Matsuoka M, Takeuchi M, Zhang J, Horiuchi Y, Anpo M, et al. Understanding TiO_2 photocatalysis: Mechanisms and materials. Chemical Reviews. 2014;**114**:9919-9986. DOI: 10.1021/cr5001892

[3] Regazzoni AE, Mandelbaum P, Matsuyoshi M, Schiller S, Bilmes SA, Blesa MA. Adsorption and photooxidation of salicylic acid on titanium dioxide: A surface complexation description. Langmuir. 1998;**14**:868-874. DOI: 10.1021/la970665n

[4] Lee D-W, Park S-J, Ihm S-K, Lee K-H. Synthesis of bimodal mesoporous titania with high thermal stability via replication of citric acid-templated mesoporous silica. Chemistry of Materials. 2007;**19**:937-941. DOI: 10.1021/cm062465f

[5] Fuentes KM, Betancourt P, Marrero S, García S. Photocatalytic degradation of phenol using doped titania supported on photonic SiO_2 spheres. Reaction Kinetics, Mechanisms and Catalysis. 2017;**120**:403-415. DOI:10.1007/s11144-016-1097-3

[6] Ullah S, Ferreira-Neto EP, Pasa AA, Alcântara CCJ, Acuña JJS, Bilmes SA, et al. Enhanced photocatalytic properties of core@shell SiO_2@TiO_2 nanoparticles. Applied Catalysis B: Environmental. 2015;**179**:333-343. DOI: 10.1016/j.apcatb.2015.05.036

[7] Chen J, Qiu F, Xu W, Cao S, Zhu H. Recent progress in enhancing photocatalytic efficiency of TiO_2-based materials. Applied Catalysis A: General. 2015;**495**:131-140. DOI: 10.1016/j.apcata.2015.02.013

[8] Zhang F, Zheng Y, Cao Y, Chen C, Zhan Y, Lin X, et al. Ordered mesoporous Ag–TiO_2–KIT-6 heterostructure: Synthesis, characterization and photocatalysis. Journal of Materials Chemistry. 2009;**19**:2771. DOI: 10.1039/b818495j

[9] Li Y, Li N, Tu J, Li X, Wang B, Chi Y, et al. TiO_2 supported on rod-like mesoporous silica SBA-15: Preparation, characterization and photocatalytic behaviour. Materials Research Bulletin. 2011;**46**:2317-2322. DOI: 10.1016/j.materresbull.2011.08.044

[10] Tian L, Liu H, Gao Y. Degradation and adsorption of rhodamine B and phenol on TiO_2/MCM-41. Kinetics and Catalysis. 2012;**53**:554-559. DOI: 10.1134/S0023158412050175

[11] Mureseanu M, Parvulescu V, Radu T, Filip M, Carja G. Mesoporous CeTiSiMCM-48 as novel photocatalyst for degradation of organic compounds. Journal of Alloys and Compounds. 2015;**648**:864-873. DOI: 10.1016/j.jallcom.2015.07.078

[12] Garg S, Narayan R, Nayak UY, Raichur AM. Mesoporous silica nanoparticles: A comprehensive review on synthesis and recent advances. Pharmaceutics. 2018;**10**:1-50. DOI: 10.3390/pharmaceutics10030118

[13] Fei X, Dong Y, Liu Y. Study on synthesis and properties of composite mesoporous TiO_2/MCM-41 photocatalysts. Materials and Technologies. 2016;**31**:423-429. DOI: 10.1179/1753555715Y.0000000073

[14] Su B-L, Sanchez C, Yang X-Y, editors. Hierarchically Structured Porous Materials. Weinheim, Germany: Wiley-VCH Verlag GmbH & Co. KGaA; 2011. DOI: 10.1002/9783527639588

[15] Sun M-H, Huang S-Z, Chen L-H, Li Y, Yang X-Y, Yuan Z-Y, et al. Applications of hierarchically structured porous materials from energy storage and conversion, catalysis, photocatalysis, adsorption, separation, and sensing to biomedicine. Chemical Society Reviews. 2016;**45**:3479-3563. DOI: 10.1039/C6CS00135A

[16] Yokoi T, Karouji T, Ohta S, Kondo JN, Tatsumi T. Synthesis of mesoporous silica nanospheres promoted by basic amino acids and their catalytic application. Chemistry of Materials. 2010;**22**:3900-3908. DOI: 10.1021/cm9037846

[17] Sun B, Zhou G, Zhang H. Synthesis, functionalization, and applications of morphology-controllable silica-based nanostructures: A review. Progress in Solid State Chemistry. 2016;**44**:1-19. DOI: 10.1016/j.progsolidstchem.2016.01.001

[18] Gustafsson H, Isaksson S, Altskär A, Holmberg K. Mesoporous silica nanoparticles with controllable morphology prepared from oil-in-water emulsions. Journal of Colloid and Interface Science. 2016;**467**:253-260. DOI: 10.1016/j.jcis.2016.01.026

[19] Maity A, Polshettiwar V. Scalable and sustainable synthesis of size-controlled monodisperse dendritic fibrous nanosilica quantified by E-factor. ACS Applied Nano Materials. 2018;**1**:3636-3643. DOI: 10.1021/acsanm.8b00761

[20] Moon D-S, Lee J-K. Tunable synthesis of hierarchical mesoporous silica nanoparticles with radial wrinkle structure. Langmuir. 2012;**28**:12341-12347. DOI: 10.1021/la302145j

[21] Yu Y-J, Xing J-L, Pang J-L, Jiang S-H, Lam K-F, Yang T-Q, et al. Facile synthesis of size controllable dendritic mesoporous silica nanoparticles. ACS Applied Materials & Interfaces. 2014;**6**:22655-22665. DOI: 10.1021/am506653n

[22] Maity A, Polshettiwar V. Dendritic fibrous nanosilica for catalysis, energy harvesting, carbon dioxide mitigation, drug delivery, and sensing. ChemSusChem. 2017;**10**:3866-3913. DOI: 10.1002/cssc.201701076

[23] Singh R, Bapat R, Qin L, Feng H, Polshettiwar V. Atomic layer deposited (ALD) TiO_2 on fibrous nano-silica (KCC-1) for photocatalysis: Nanoparticle formation and size quantization effect. ACS Catalysis. 2016;**6**:2770-2784. DOI: 10.1021/acscatal.6b00418

[24] Schacht S, Huo Q, Voigt-Martin IG, Stucky GD, Schuth F. Oil-water interface templating of mesoporous macroscale structures. Science (80-.). 1996;**273**:768-771. DOI: 10.1126/science.273.5276.768

[25] Sanchez-Dominguez M, Pemartin K, Boutonnet M. Preparation of inorganic nanoparticles in oil-in-water microemulsions: A soft and versatile approach. Current Opinion in Colloid & Interface Science. 2012;**17**:297-305. DOI: 10.1016/j.cocis.2012.06.007

[26] Kuijk A, van Blaaderen A, Imhof A. Synthesis of monodisperse, rodlike silica colloids with tunable aspect ratio. Journal of the American Chemical Society. 2011;**133**:2346-2349. DOI: 10.1021/ja109524h

[27] Lind A, Spliethoff B, Lindén M. Unusual, vesicle-like patterned, mesoscopically ordered silica. Chemistry of Materials. 2003;**15**:813-818. DOI: 10.1021/cm021243o

[28] Wang S, Chen M, Wu L. One-step synthesis of cagelike hollow silica spheres with large through-holes for macromolecule delivery. ACS Applied Materials & Interfaces. 2016;**8**:33316-33325. DOI: 10.1021/acsami.6b11639

[29] Polshettiwar V, Cha D, Zhang X, Basset JM. High-surface-area silica nanospheres (KCC-1) with a fibrous morphology. Angewandte Chemie International Edition. 2010;**49**:9652-9656. DOI: 10.1002/anie.201003451

[30] ALOthman Z. A review: Fundamental aspects of silicate mesoporous materials. Materials (Basel). 2012;**5**:2874-2902. DOI: 10.3390/ma5122874

[31] Putz A-M, Cecilia S, Ianăşi C, Dudás Z, Székely KN, Plocek J, et al. Pore ordering in mesoporous matrices induced by different directing agents. Journal of Porous Materials. 2015;**22**:321-331. DOI: 10.1007/s10934-014-9899-z

[32] Moon D-S, Lee J-K. Formation of wrinkled silica mesostructures based on the phase behavior of pseudoternary systems. Langmuir. 2014;**30**:15574-15580. DOI: 10.1021/la504207k

[33] Rana S, Bhattacharjee J, Barick KC, Verma G, Hassan PA, Yakhmi JV. Interfacial engineering of nanoparticles for cancer therapeutics. In: Nanostructures Cancer Therapy. Amsterdam: Elsevier; 2017. pp. 177-209. DOI: 10.1016/B978-0-323-46144-3.00007-6

[34] Pileni MP. Reverse micelles as microreactors. The Journal of Physical Chemistry. 1993;**97**:6961-6973. DOI: 10.1021/j100129a008

[35] Danielsson I, Lindman B. The definition of microemulsion. Colloids and Surfaces. 1981;**3**:391-392. DOI: 10.1016/0166-6622(81)80064-9

[36] Aubery C, Solans C, Prevost S, Gradzielski M, Sanchez-Dominguez M. Microemulsions as reaction media for the synthesis of mixed oxide nanoparticles: Relationships between microemulsion structure, reactivity, and nanoparticle characteristics. Langmuir. 2013;**29**:1779-1789. DOI: 10.1021/la303817w

[37] Nič M, Jirát J, Košata B, Jenkins A, McNaught A, editors. IUPAC Compendium of Chemical Terminology. Research Triagle Park, NC: IUPAC; 2009. DOI: 10.1351/goldbook

[38] Hench LL, West JK. The sol-gel process. Chemical Reviews. 1990;**90**:33-72. DOI: 10.1021/cr00099a003

[39] Brinker CJ. Hydrolysis and condensation of silicates: Effects on structure. Journal of Non-Crystalline Solids. 1988;**100**:31-50. DOI: 10.1016/0022-3093(88)90005-1

[40] Febriyanti E, Suendo V, Mukti RR, Prasetyo A, Arifin AF, Akbar MA, et al. Ismunandar, further insight into the definite morphology and formation mechanism of mesoporous silica KCC-1. Langmuir. 2016;**32**:5802-5811. DOI: 10.1021/acs.langmuir.6b00675

[41] Maity A, Das A, Sen D, Mazumder S, Polshettiwar V. Unraveling the formation mechanism of dendritic fibrous nanosilica. Langmuir. 2017;**33**:13774-13782. DOI: 10.1021/acs.langmuir.7b02996

[42] Bayal N, Singh B, Singh R, Polshettiwar V. Size and fiber density controlled synthesis of fibrous nanosilica spheres (KCC-1). Scientific Reports. 2016;**6**:24888. DOI: 10.1038/srep24888

[43] Shen D, Yang J, Li X, Zhou L, Zhang R, Li W, et al. Biphase stratification approach to three-dimensional dendritic biodegradable mesoporous silica nanospheres. Nano Letters. 2014;**14**:923-932. DOI: 10.1021/nl404316v

[44] Yamada H, Urata C, Higashitamori S, Aoyama Y, Yamauchi Y, Kuroda K. Critical roles of cationic surfactants in the preparation of colloidal mesostructured silica nanoparticles: Control of mesostructure, particle size,

and dispersion. ACS Applied Materials & Interfaces. 2014;**6**:3491-3500. DOI: 10.1021/am405633r

[45] Yokoi T, Yoshitake H, Tatsumi T. Synthesis of amino-functionalized MCM-41 via direct co-condensation and post-synthesis grafting methods using mono-, di- and tri-amino-organoalkoxysilanes. Electronic supplementary information (ESI) available: XRD patterns of (A) x-dNN-MCM-41 and (B) x-dNNN. Journal of Materials Chemistry. 2004;**14**:951. DOI: 10.1039/b310576h

[46] Clark KG, Gaddy VL, Rist CE. Equilibria in the ammonium carbamate-urea-water system. Industrial and Engineering Chemistry. 1933;**25**:1092-1096. DOI: 10.1021/ie50286a008

[47] Jin H, Liu Z, Ohsuna T, Terasaki O, Inoue Y, Sakamoto K, et al. Control of morphology and helicity of chiral mesoporous silica. Advanced Materials. 2006;**18**:593-596. DOI: 10.1002/adma.200502038

[48] Hu Y, Martra G, Zhang J, Higashimoto S, Coluccia S, Anpo M. Characterization of the local structures of Ti-MCM-41 and their photocatalytic reactivity for the decomposition of NO into N_2 and O_2. The Journal of Physical Chemistry. B. 2006;**110**:1680-1685. DOI: 10.1021/jp058240u

[49] Li J, Zhou C, Xie H, Ge Z, Yuan L, Li X. Titanium-containing mesoporous materials: Synthesis and application in selective catalytic oxidation. Journal of Natural Gas Chemistry. 2006;**15**:164-177. DOI: 10.1016/S1003-9953(06)60023-6

[50] Livage J, Sanchez C. Sol-gel chemistry. Journal of Non-Crystalline Solids. 1992;**145**:11-19. DOI: 10.1016/S0022-3093(05)80422-3

[51] Angelomé PC, Andrini L, Calvo ME, Requejo FG, Bilmes SA, Soler-Illia GJAA. Mesoporous anatase TiO_2 films: Use of Ti K XANES for the quantification of the nanocrystalline character and substrate effects in the photocatalysis behavior. Journal of Physical Chemistry C. 2007;**111**: 10886-10893. DOI: 10.1021/jp069020z

[52] Zhuravlev LT. Concentration of hydroxyl groups on the surface of amorphous silicas. Langmuir. 1987;**3**:316-318

[53] Kamegawa T, Yamashita H. Photocatalytic properties of TiO_2-loaded porous silica with hierarchical macroporous and mesoporous architectures. Nanostructures Science and Technology. Switzerland: Springer International Publishing; In: 2016: pp. 229-240. DOI: 10.1007/978-3-319-26079-2_13

[54] Ding Z, Lu GQ, Greenfield PF. Role of the crystallite phase of TiO_2 in heterogeneous photocatalysis for phenol oxidation in water. The Journal of Physical Chemistry. B. 2000;**104**: 4815-4820. DOI: 10.1021/jp993819b

[55] Hanaor DAH, Sorrell CC. Review of the anatase to rutile phase transformation. Journal of Materials Science. 2011;**46**:855-874. DOI: 10.1007/s10853-010-5113-0

[56] Kundu S, Polshettiwar V. Hydrothermal crystallization of nano-titanium dioxide for enhanced photocatalytic hydrogen generation. ChemPhotoChem. 2018;**2**:796-800 (Copyright Wiley-VCH Verlag GmbH & Co. KGaA. Reproduced with permission. DOI: 10.1002/cptc.201800101

Chapter 4

Toward the Creation of Highly Active Photocatalysts That Convert Methane into Methanol

Yuichi Negishi, Seiichiro Watanabe, Marika Aoki, Sakiat Hossain and Wataru Kurashige

Abstract

Methane exists abundantly around Japan as methane hydrate. As the effective use of such methane, the conversion of methane into methanol has recently attracted much attention. Photocatalytic reaction is one of the methods which convert methane into methanol without using much energy. However, it is indispensable to improve the photocatalytic activity for their practical use. Our group has attempted to improve the activity of mesoporous tungsten trioxide and titanium dioxide (m-WO_3 and m-TiO_2) photocatalysts, which convert methane into methanol, by loading the ultrafine metal clusters as cocatalyst on the photocatalysts. As a result, we have succeeded in loading ultrafine metal-cluster cocatalysts onto m-WO_3 and m-TiO_2 and thereby improving their photocatalytic activity. Our study also demonstrated that the kind of metal element suitable for each photocatalyst depends on the kind of the photocatalysts, and thereby it is important to select the metal clusters suitable for each photocatalyst for improving its photocatalytic activity.

Keywords: photocatalyst, methane, methanol, activation, cocatalyst

1. Introduction

In recent years, the effective use of methane (CH_4) has attracted attention owing to the large amount of methane hydrate which is estimated to exist under the sea around Japan (**Figure 1(a)**) [1]. When CH_4 is burnt for generating energy, it generates small amounts of compounds which cause environmental problems, such as carbon dioxide, nitrogen oxide, and sulfur oxide (**Figure 1(b)**), [2]. Therefore, CH_4 can be used as a comparatively clean energy source. However, since CH_4 is a gas state at room temperature, it occupies a large volume and therefore costs for the transportation. On the other hand, if CH_4 is converted to liquid methanol (CH_3OH), the transportation costs could be reduced. Furthermore, the generated CH_3OH could be effectively used as a raw material for producing various chemical compounds. Therefore, in recent years, considerable attention has been focused on the development of a methodology for efficiently converting CH_4 into CH_3OH.

Regarding such a conversion, the current main methods require extreme conditions such as high temperature of around 500°C and high pressure of 50 atm or more [3]. Furthermore, since the reaction proceeds in two stages, the conversion consumes a large amount of energy. Therefore, many studies have been conducted to find a

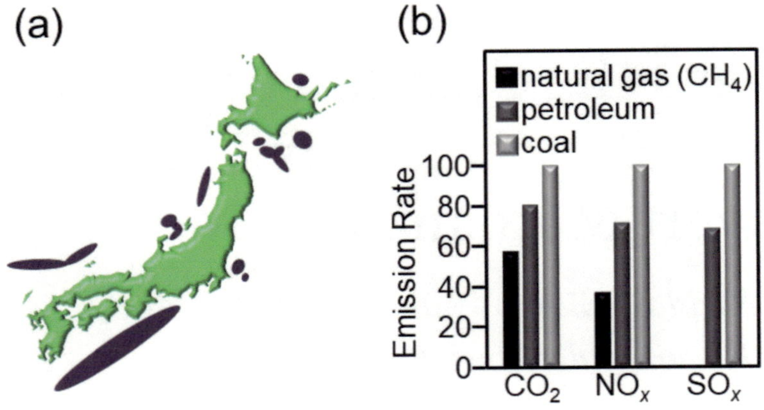

Figure 1.
(a) Estimated distribution of methane hydrate lying under the sea around Japan. (b) emission rate of each gas (CO_2, NO_x, or SO_x) from CH_4, petroleum, and coal. In (b), emission rate of each gas is normalized by letting the emission rate of CO_2 from coal be 100. Adapted from [1, 2].

method for the direct conversion of CH_4 to CH_3OH under mild conditions. However, most direct reactions require also high temperature and pressure conditions together with the use of an oxidizing agent [4]. Using a photocatalytic reaction for this conversion, the reaction can be conducted under moderate conditions [5]. However, the catalytic activity is still low, and further improvement is required for practical application. Then, our group has attempted to improve the photocatalytic activity of the photocatalyst used in this conversion. Our strategy was to load the controlled metal clusters on the surface of the photocatalyst as active sites (cocatalyst).

2. Research examples

Both tungsten oxide (WO_3) and titanium oxide (TiO_2) have redox potentials that are sufficient to generate hydroxyl radicals (·OH) (**Figure 2(a)**), which is necessary for the reaction to proceed (**Figure 2(b)**) [6]. Furthermore, in their mesoporous structures (m-WO_3 and m-TiO_2), surface area becomes large and thereby the adsorption of reactive molecules is enhanced. In addition, in m-WO_3 and m-TiO_2, since the regular arrangement of pores suppresses the recombination of electrons and holes generated by the light irradiation, electrons and holes can be efficiently used in the reaction [7]. Then, in our study, we attempted to activate such m-WO_3 and m-TiO_2 by loading the ultrafine metal clusters as cocatalyst on the photocatalysts.

2.1 m-WO_3 photocatalyst

2.1.1 Preparation and loading of cocatalyst

m-WO_3 was prepared using a hard template method in the same manner as that reported in the literature [8]. First, mesoporous silica (KIT-6) was prepared to be used as a template (**Figure 3(a)**). Then, the obtained KIT-6 was mixed with 12-tungsto(VI) phosphoric acid *n*-hydrate ($H_3(PW_{12}O_{40})\cdot nH_2O$) in ethanol. The product was dried and then calcined at 350°C for 4 h. To the obtained powder, a further $H_3(PW_{12}O_{40})\cdot nH_2O$ ethanol solution was added and the mixture was stirred for 30 min. The product was again dried and then calcined at 550°C for 6 h. Finally, m-WO_3 was obtained by removing the KIT-6 template using hydrofluoric acid (HF) (**Figure 3(b)**).

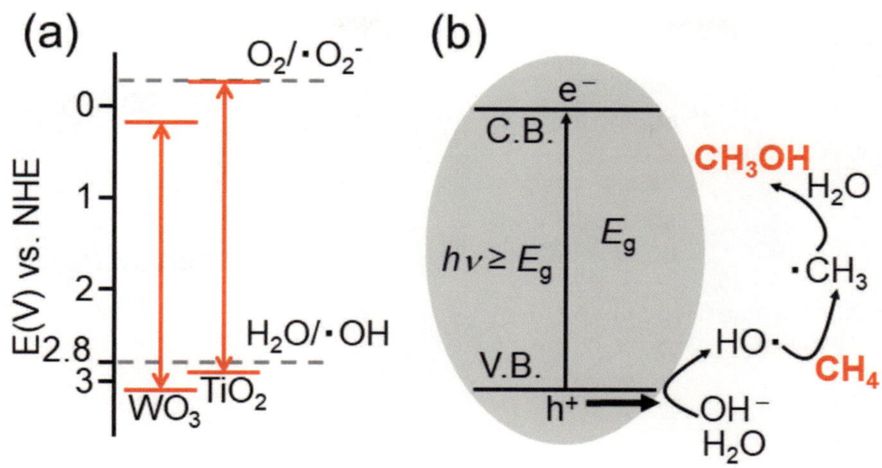

Figure 2.
(a) Redox potentials of WO3 and TiO_2. Adapted from [4]. (b) Mechanism of the conversion of CH_4 to CH_3OH by photocatalytic reaction; C.B., V.B., and E_g indicate conduction band, valence band, and band gap, respectively.

Next, metal clusters were loaded onto the obtained m-WO_3. Our previous research on water-splitting photocatalysts [9–13] revealed that when a chemically-synthesized ligand-protected metal cluster is adsorbed onto a photocatalyst and its ligand is removed by the calcination, the controlled metal clusters can be loaded on the photocatalyst. Then, in this study, each metal cluster was loaded onto m-WO_3 using such a method (**Figure 3(c)**). Silver (Ag), nickel (Ni), and cobalt (Co) were chosen as metal element because they were expected to work as cocatalyst [14, 15]. Glutathionate (SG) was used as the ligand of the metal cluster. SG-protected silver clusters ($Ag_n(SG)_m$) were synthesized using the solid phase method reported by Bakr *et al* [16]. $Ni_n(SG)_m$ and $Co_n(SG)_m$ clusters were synthesized using the liquid phase method reported for SG-protected gold clusters ($Au_n(SG)_m$) [17]. The $M_n(SG)_m$ clusters (M = Ag, Ni, or Co) thus obtained were dissolved in water, and m-WO_3 was added to this solution to adsorb the $M_n(SG)_m$ clusters on m-WO_3 ($M_n(SG)_m$-m-WO_3; M = Ag, Ni, or Co). The obtained $M_n(SG)_m$-m-WO_3 was calcined at 500°C under atmospheric pressure to remove the ligand of the $M_n(SG)_m$ cluster, and thereby, each metal cluster was loaded on m-WO_3 (M_n-m-WO_3; M = Ag, Ni, or Co).

2.1.2 Structural characterization

In the transmission electron microscope (TEM) image of m-WO_3, pores of 10 nm or less were observed (**Figure 4(a,b)**). In the small angle X-ray diffraction (XRD) pattern of m-WO_3 (**Figure 4(c)**), similar to the literature [18], a peak assigned to the (211) plane of KIT-6 was observed, in addition to the peak attributed to the (110) plane derived from cubic $I4_132$ or $I4_332$. These results indicate that the synthesized m-WO_3 has an arranged structure with regular pores.

Figure 5(a,b) shows the TEM images of $Co_n(SG)_m$-m-WO_3 and Co_n-m-WO_3, respectively, as representative examples of photocatalysts after adsorption and calcination. Particles with a size of 0.93 ± 0.20 nm could be observed in the TEM image of $Co_n(SG)_m$-m-WO_3. This indicates that ultrafine $Co_n(SG)_m$ clusters were synthesized with a narrow distribution. Particles of a similar size (0.96 ± 0.19 nm) were also observed for Co_n-m-WO_3. This indicates that Co_n clusters were loaded onto m-WO_3 during calcination without the aggregation. Based on the bulk density of Co (8.900 g/cm^3), it can be estimated that each loaded particle contains about 40 Co atoms. The loading of ultrafine particles was similarly observed in other

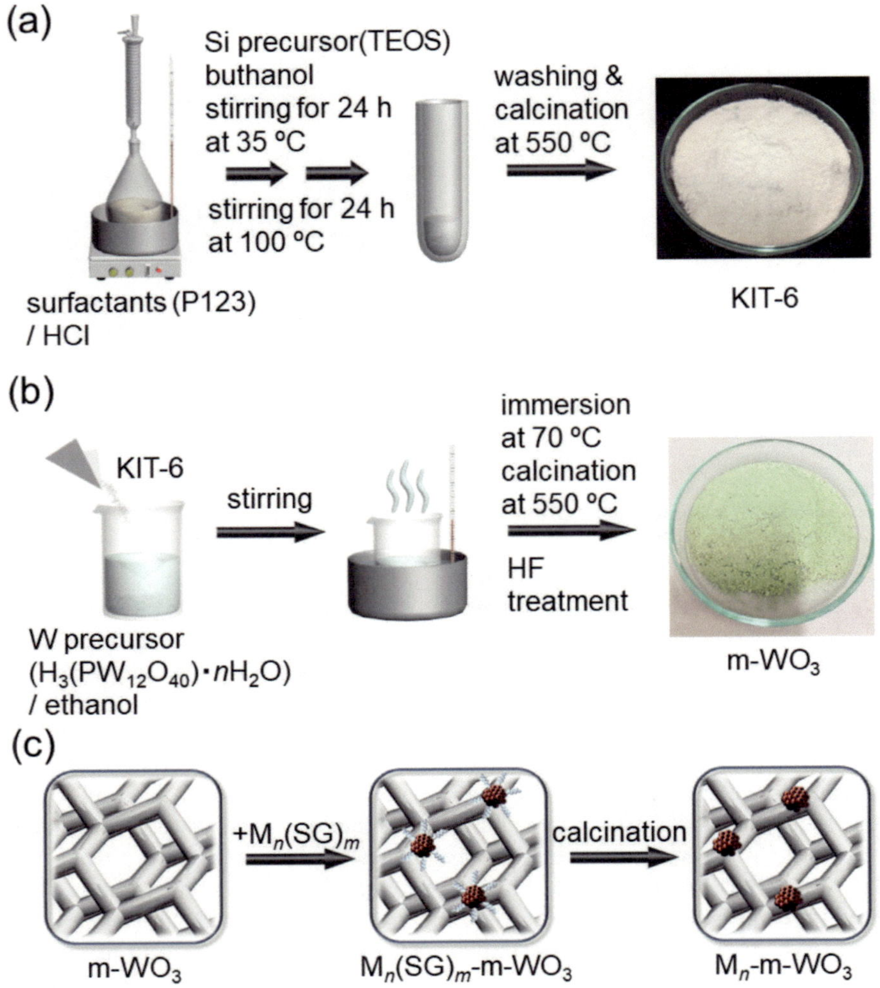

Figure 3.
Preparation methods of (a) KIT-6, (b) m-WO_3, and (c) M_n-m-WO_3 (M = Ag, Ni, or Co).

M_n-m-WO_3 (M = Ag or Ni). In this way, we have succeeded in loading ultrafine M_n clusters on m-WO_3 with a narrow size distribution.

2.1.3 Evaluation of photocatalytic activity

The photocatalytic activity was measured using an experimental apparatus built in-house consisting of a high-pressure mercury (Hg) lamp (400 W) and a quartz cell (**Figure 6**) [11]. First, 300 mg of M_n-m-WO_3 (M = Ag, Ni, and Co) was dispersed in 300 mL of water. Then, CH_4 and helium carrier gas were passed through this solution at a flow rate of 4.5–5 mL/min and 18 mL/min, respectively. The reaction was performed by the irradiation of ultraviolet light using a high-pressure Hg lamp. The temperature of the reaction system was maintained at 55–60°C by a circulator, and the gas generated by the reaction was analyzed with gas chromatography.

Figure 7 shows the rate of CH_3OH evolution from an aqueous solution containing M_n-m-WO_3 or m-WO_3. All the photocatalysts loaded with M_n clusters evolved CH_3OH at faster rate as compared with m-WO_3 without cocatalyst. This indicates

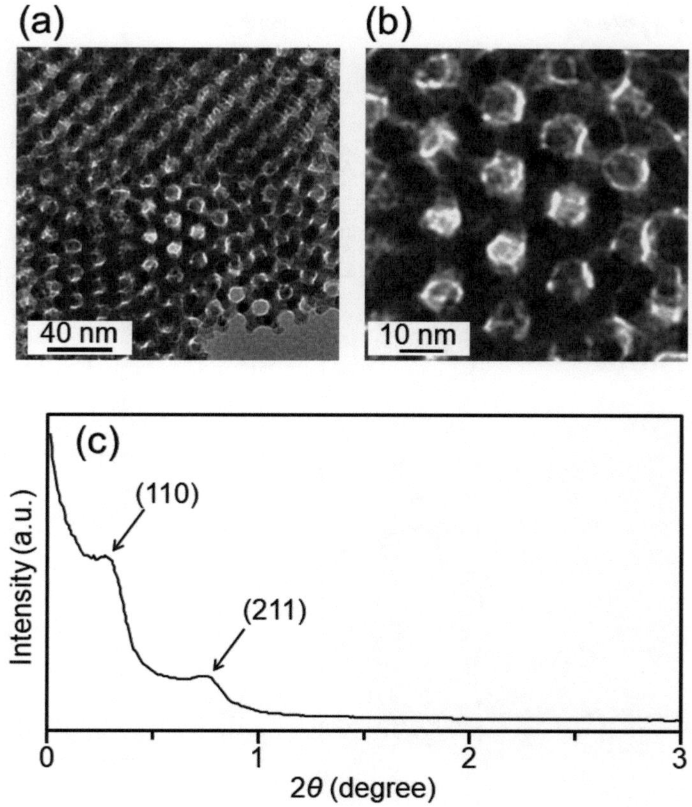

Figure 4.
Characterization of m-WO₃. (a) TEM image, (b) expanded TEM image, and (c) small angle XRD pattern.

that the loading of the M_n clusters improved the rate of the conversion of CH_4 to CH_3OH. It was found that the Co_n clusters have the largest improvement effect among the M_n clusters used in this study.

According to a previous study on m-WO_3 [8], in the conversion of CH_4 to CH_3OH, first, the holes generated via photoexcitation oxidize the hydroxyl groups (OH^-) on the photocatalyst surface and/or water (H_2O_{ad}) adsorbed on the surface to form ·OH radicals. The ·OH radicals have intense oxidizing power and attack CH_4 to generate methyl radicals (·CH_3). The ·CH_3 radicals thus produced react with H_2O to produce CH_3OH (**Figure 2(b)**). It can be considered that loading M_n clusters onto m-WO_3 promotes the charge separation between the electrons and holes, and thereby, the holes are efficiently consumed [15], leading to the improvement of the activity. It is presumed that highest activity was observed in the reaction using Co_n-m-WO_3, since the relationship between the positions of valence band of the photocatalyst and the orbitals of the loaded M_n cluster and/or the relationship between the redox potential of ·OH radical and that of the loaded M_n cluster was effective for accelerating the reaction in Co_n-m-WO_3 compared with the other photocatalysts (Ag_n-m-WO_3 or Ni_n-m-WO_3).

2.2 m-TiO_2 photocatalyst

2.2.1 Preparation and loading of cocatalyst

A m-TiO_2 photocatalyst was prepared according to a method reported in the literature [19]. Specifically, m-TiO_2 photocatalyst was prepared by forming a

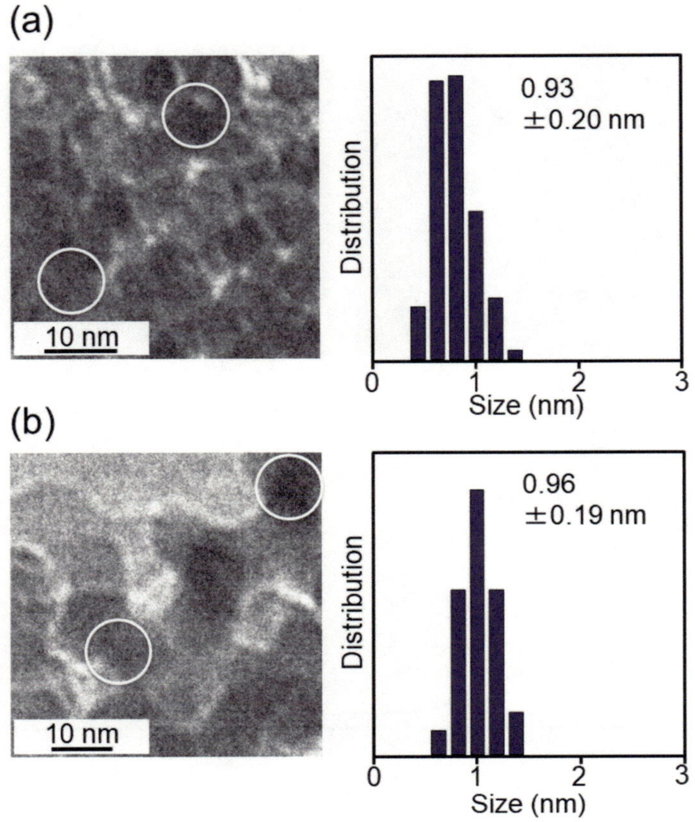

Figure 5.
TEM image and particle size distribution of (a) $Co_n(SG)_m$-m-WO_3 and (b) Co_n-m-WO_3. Multiple ultrafine particles can be seen in the white circles.

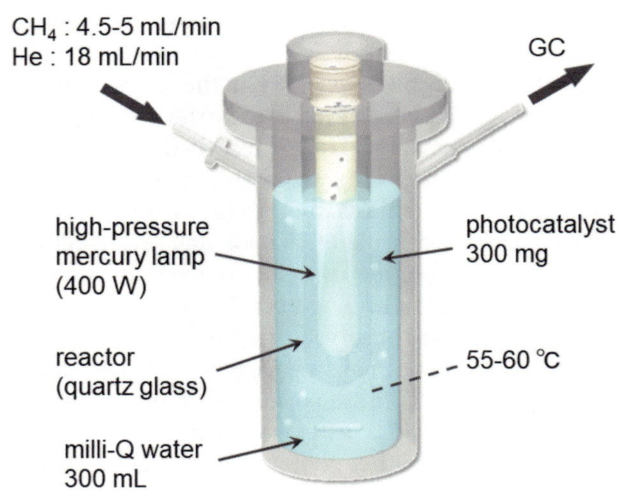

Figure 6.
Experimental apparatus built in-house consisting of a high-pressure mercury (Hg) lamp (400 W) and a quartz cell. GC means gas chromatograph.

photocatalyst around the micelle composed of surfactants. First, hexadecyltrimethylammonium bromide ($[(C_{16}H_{33})N(CH_3)_3]Br$) was added to an aqueous solution of titanyl oxysulfate sulfuric acid hydrate ($TiOSO_4$), and the solution

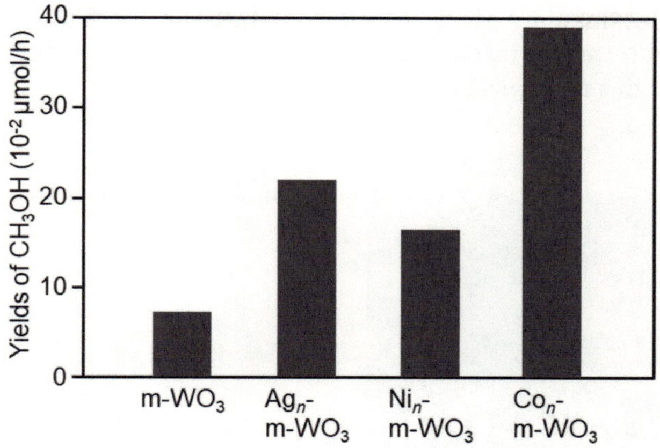

Figure 7.
Rate of CH_3OH evolution from an aqueous solution containing M_n-m-WO_3 or m-WO_3.

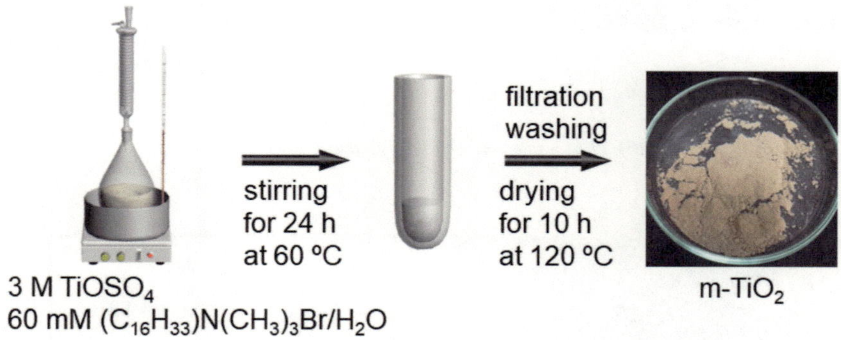

Figure 8.
Preparation methods of m-TiO_2.

was stirred at 60°C for 24 h. The obtained precipitate was washed with water and dried at 120°C for 10 h. Then, the dried sample was calcined at 450°C for 2 h to obtain m-TiO_2 (**Figure 8**). The M_n clusters (M = Ag, Ni, or Co) were loaded onto the surface of the obtained m-TiO_2 using the same method as that described for M_n-m-WO_3 (**Figure 3(c)**).

2.2.2 Structural characterization

Figure 9(a) shows a TEM image of the prepared m-TiO_2. It can be confirmed from this image that the material has a mesoporous structure different from that of m-WO_3 (**Figure 4(b)**). The diffuse reflection (DR) spectrum of m-TiO_2 revealed that the obtained m-TiO_2 has a band gap of around 3.24 eV, indicating that the obtained m-TiO_2 has an anatase type structure [19].

Figure 10(a) shows a TEM image of the $Ni_n(SG)_m$ cluster, as a representative example of a synthesized $M_n(SG)_m$ cluster. The particles with a size of 0.70 ± 0.14 nm can be observed in **Figure 10(a)**. On the other hand, the particles with a slightly larger size of 0.92 ± 0.17 nm were observed in the TEM image of Ni_n-m-TiO_2 (**Figure 10(b)**). This indicates that, in the synthesis of $Ni_n(SG)_m$-m-TiO_2, some cluster aggregation occurred during the adsorption or calcination processes. However, the size distribution of the observed particles is very narrow even for loaded particles. The number of Ni atoms contained in each particle was estimated to be about 40, based on the bulk density of Ni (8.908 g/cm^3). Loading of such ultrafine particles was similarly observed for the other M_n-m-TiO_2 (M = Ag or Co). These results indicate that m-TiO_2 ultrafine clusters were successfully loaded on a photocatalyst with a narrow distribution of the particle size.

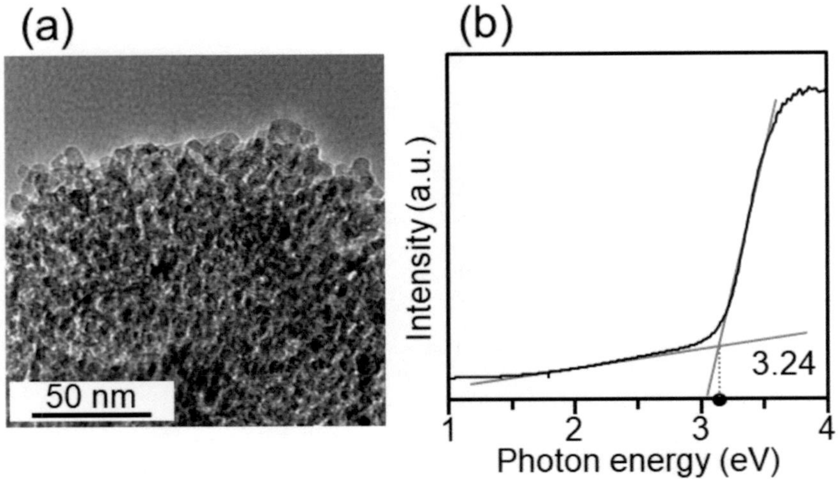

Figure 9.
Characterization of m-TiO_2. (a) TEM image and (b) DR spectrum.

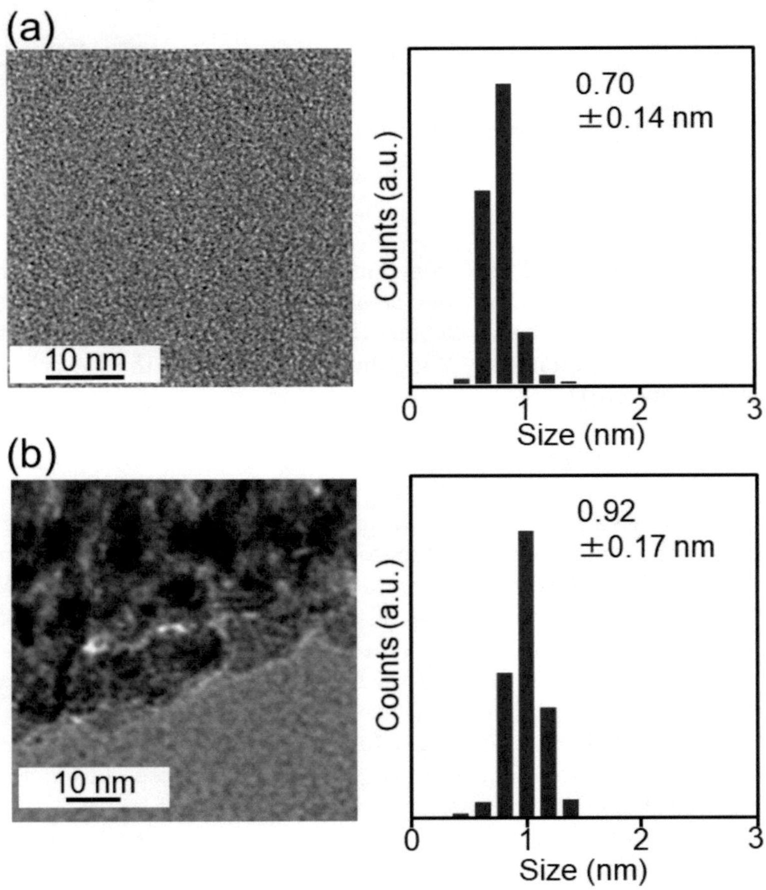

Figure 10.
TEM image and particle size distribution of (a) $Ni_n(SG)_m$ and (b) Ni_n-m-TiO_2.

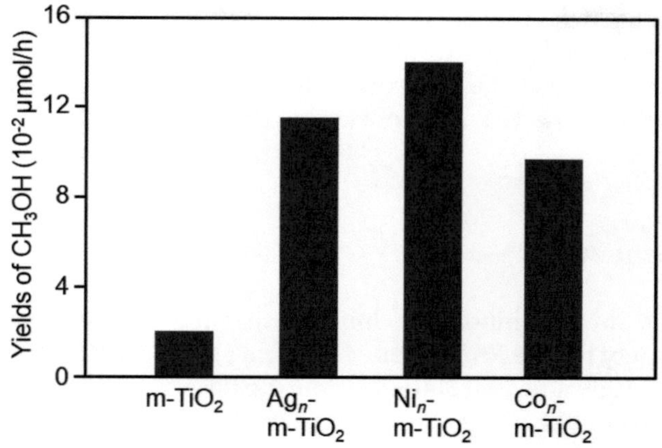

Figure 11.
Rate of CH_3OH evolution from an aqueous solution containing M_n-m-TiO_2 or m-TiO_2.

2.2.3 Evaluation of photocatalytic activity

The photocatalytic activity of a series of M_n-m-TiO_2 obtained in this manner was estimated in the same manner as that described in Section 2.1.3. **Figure 11** shows the rate of CH_3OH evolution from aqueous solutions containing either M_n-m-TiO_2 or m-TiO_2. Overall, the rate of CH_3OH evolution from M_n-m-TiO_2 or m-TiO_2 was lower than that from M_n-m-WO_3 or m-WO_3. Since the position of conduction band is relatively high in TiO_2 (**Figure 2(a)**), the superoxide radicals with strong oxidizing power can be generated from aqueous solution containing M_n-m-TiO_2 or m-TiO_2 by photoirradiation. It can be considered that since the superoxide radicals such generated attacked CH_3OH and thereby the excessive oxidation (decomposition) of CH_3OH occurred [20], the rate of the CH_3OH evolution decreased as a whole in the case of M_n-m-TiO_2 or m-TiO_2 compared to the case of M_n-m-WO_3 or m-WO_3. **Figure 11** also revealed that the loading of M_n clusters has the effect of promoting the conversion of CH_4 into CH_3OH also in the case of M_n-m-TiO_2.

On the other hand, regarding the metal elements, Ni showed a particularly high cocatalytic effect in M_n-m-TiO_2, unlike the case of m-WO_3 (**Figure 11**). The relationship between the position of the valence band of the photocatalyst and the position of the orbitals of the loaded M_n cluster and/or the relationship between the redox potential of ·OH radical and that of the loaded M_n cluster are considered to be strongly related to these phenomena. These results indicate that choosing an appropriate cocatalyst depending on the type of photocatalyst is very important to improve the photocatalytic activity.

3. Conclusions

Ultrafine M_n clusters (M = Ag, Ni, or Co) with a particle size of around 1 nm were successfully loaded onto m-WO_3 and m-TiO_2 with a narrow distribution. The photocatalytic activity measurements clearly demonstrated that the loading of such cocatalysts is effective in improving the activity for both types of photocatalysts. Furthermore, our study also revealed that it is important to appropriately selecting the element of the cocatalyst according to the photocatalyst to improve the photocatalytic activity. In these experiments, the particle diameter of the cocatalyst was reduced to approximately 1 nm to increase the reactive surface area. However, this size of a cocatalyst particle might not be optimal for promoting the reaction. In fact, our previous studies on water-splitting photocatalysts showed that extreme refining of the cocatalyst particles reduces the activity per a surface atom of cocatalysts [8]. Therefore, in future studies, we would like to clarify the correlation between the particle size and activity of the cocatalyst. Furthermore, it seems also necessary to consider the method for efficiently consuming the excited electrons to accelerate the reaction.

Acknowledgements

We thank Mr. Shun Yoshino for technical assistance. This work was supported by the Japan Society for the Promotion of Science (JSPS) KAKENHI (grant numbers JP16H04099 and 16 K21402), Scientific Research on Innovative Areas "Coordination Asymmetry" (grant number 17H05385), and Scientific Research on Innovative Areas "Innovations for Light-Energy Conversion" (grant number 18H05178). Funding from the Takahashi Industrial and Economic Research Foundation, Futaba Electronics Memorial Foundation, Iwatani Naoji Foundation, and Asahi Glass Foundation is also gratefully acknowledged.

Author details

Yuichi Negishi[1,2*], Seiichiro Watanabe[1], Marika Aoki[1], Sakiat Hossain[1] and Wataru Kurashige[1]

1 Department of Applied Chemistry, Faculty of Science, Tokyo University of Science, Tokyo, Japan

2 Photocatalysis International Research Center, Tokyo University of Science, Chiba, Japan

*Address all correspondence to: negishi@rs.kagu.tus.ac.jp

References

[1] Tsuji Y, Ishida H, Nakamizu M, Matsumoto R, Shimizu S. Overview of the MITI nankai trough wells: A milestone in the evaluation of methane hydrate resources. Resource Geology. 2004;**54**:3-10. DOI: 10.1111/j.1751-3928.2004.tb00182.x

[2] Pascoli SD, Femia A, Luzzati T. Natural gas, cars and the environment. A (relatively) 'clean' and cheap fuel looking for users. Ecological Economics. 2001;**38**:179-189. DOI: 10.1016/S0921-8009(01)00174-4

[3] Sushkevich VL, Palagin D, Ranocchiari M, Bokhoven JAV. Selective anaerobic oxidation of methane enables direct synthesis of methanol. Science. 2017;**356**:523-527. DOI: 10.1126/science.aam9035

[4] Zakaria Z, Kamarudin SK. Direct conversion technologies of methane to methanol: An overview. Renewable and Sustainable Energy Reviews. 2016;**65**:250-261. DOI: 10.1016/j.rser.2016.05.082

[5] Zhu W, Shen M, Fan G, Yang A, Meyer JR, Ou Y, et al. Facet-dependent enhancement in the activity of bismuth vanadate microcrystals for the photocatalytic conversion of methane to methanol. ACS Applied Nano Materials. 2018;**1**:6683-6691. DOI: 10.1021/acsanm.8b01490

[6] Murcia-López SM, Villa K, Andreu T, Morante JR. Partial oxidation of methane to methanol using bismuth-Based photocatalysts. ACS Catalysis. 2014;**4**:3013-3019. DOI: 10.1021/cs500821r

[7] Li L, Krissanasaeranee M, Pattinson SW, Stefik M, Wiesner U, Steiner U, et al. Enhanced photocatalytic properties in well-ordered mesoporous WO_3. Chemical Communications. 2010;**46**:7620-7622. DOI: 10.1039/C0CC01237H

[8] Villa K, Murcia-López S, Andreu T, Morante JR. Mesoporous WO_3 photocatalyst for the partial oxidation of methane to methanol using electron scavengers. Applied Catalysis B: Environmental. 2015;**163**:150-155. DOI: 10.1016/j.apcatb.2014.07.055

[9] Negishi Y, Mizuno M, Hirayama M, Omatoi M, Takayama T, Iwase A, et al. Enhanced photocatalytic water splitting by $BaLa_4Ti_4O_{15}$ loaded with ~1 nm gold nanoclusters using glutathione-protected Au_{25} clusters. Nanoscale. 2013;**5**:7188-7192. DOI: 10.1039/c3nr01888A

[10] Negishi Y, Matsuura Y, Tomizawa R, Kurashige W, Niihori Y, Takayama T, et al. Controlled loading of small Au_n clusters (n = 10-39) onto $BaLa_4Ti_4O_{15}$ photocatalysts: Toward an understanding of size effect of cocatalyst on water-splitting photocatalytic activity. The Journal of Physical Chemistry C. 2015;**119**:11224-11232. DOI: 10.1021/jp5122432

[11] Kurashige W, Kumazawa R, Ishii D, Hayashi R, Niihori Y, Hossain S, et al. Au_{25}-loaded $BaLa_4Ti_4O_{15}$ water-splitting photocatalyst with enhanced activity and durability produced using new chromium oxide shell formation method. The Journal of Physical Chemistry C. 2018;**122**:13669-13681. DOI: 10.1021/acs.jpcc.8b00151

[12] Negishi Y. Toward the creation of functionalized metal nanoclusters and highly active photocatalytic materials using thiolate-protected magic gold clusters. Bulletin of the Chemical Society of Japan. 2014;**87**:375-389. DOI: 10.1246/bcsj.20130288

[13] Kurashige W, Niihori Y, Sharma S, Negishi Y. Precise synthesis, functionalization and application of thiolate-protected gold clusters. Coordination Chemistry Reviews.

2016;**320-321**:238-250. DOI: 10.1016/j.ccr.2016.02.013

[14] Hameed A, Ismail IMI, Aslam M, Gondal MA. Photocatalytic conversion of methane into methanol: Performance of silver impregnated WO_3. Applied Catalysis A: General. 2014;**470**:327-335. DOI: 10.1016/j.apcata.2013.10.045

[15] Liu L, Ji Z, Zou W, Gu X, Deng Y, Gao F, et al. In situ loading transition metal oxide clusters on TiO_2 Nanosheets as Co-catalysts for exceptional high photoactivity. ACS Catalysis. 2013;**3**:2052-2061. DOI: 10.1021/cs4002755

[16] Bootharaju MS, Burlakov VM, Besong TMD, Joshi CP, AbdulHalim LG, Black DM, et al. Reversible size control of silver nanoclusters via ligand-Exchange. Chemistry of Materials. 2015;**27**:4289-4297. DOI: 10.1021/acs.chemmater.5b00650

[17] Negishi Y, Nobusada K, Tsukuda T. Glutathione-protected gold clusters revisited: bridging the gap between gold(I)–thiolate complexes and thiolate-protected gold nanocrystals. Journal of the American Chemical Society. 2005;**127**:5261-5270. DOI: 10.1021/ja042218h

[18] Rossinyol E, Arbiol J, Peiró F, Cornet A, Morante JR, Tian B, et al. Nanostructured metal oxides synthesized by hard template method for gas sensing applications. Sensors and Actuators B: Chemical. 2005;**109**:57-63. DOI: 10.1016/j.snb.2005.03.016

[19] Shibata H, Ogura T, Mukai T, Ohkubo T, Sakai H, Abe M. Direct synthesis of mesoporous titania particles having a crystalline wall. Journal of the American Chemical Society. 2005;**127**:16396-16397. DOI: 10.1021/ja0552601

[20] Gondal MA, Hameed A, Yamani ZH, Arfaj A. Photocatalytic transformation of methane into methanol under UV laser irradiation over WO_3, TiO_2 and NiO catalysts. Chemical Physics Letters. 2004;**392**: 372-377. DOI: 10.1016/j.cplett.2004.05.0

Chapter 5

Rare-Earth-Based Materials for Heterogeneous Photocatalysis

Sahar Zinatloo-Ajabshir and Zahra Sayyar

Abstract

Recently, the synthesis of rare-earth-based nanostructures as a significant class of materials with photocatalysis activity has attracted the attention of researchers. Many studies have shown their applications in various fields, specifically in photocatalysis. There are different methods to synthesis of rare-earth nanostructures. In this study, we discuss about modification of rare-earth-based materials. Also production methods and their advantages and disadvantages have been presented, briefly. Finally, photocatalytic applications of rare-earth nanostructures are highlighted.

Keywords: rare-earth materials, heterogeneous photocatalysis, nanostructure, methods, modification

1. Introduction

Nanostructures especially rare-earth-based materials have attracted interest of researches in recent decade due to their unique and important properties such as electronic, optical, photocatalytic and magnetic. These materials with different shapes can be used in various fields because of potential applications. Because their dimension and shape the behavior of nanoscale compounds is greatly sensitive [1].

A set of chemical elements in the periodic table, specifically scandium (Sc), yttrium (Y), lanthanum (La) and all of the lanthanide series (Ln), are rare-earth materials. All rare-earth materials have unique properties corresponding to the filling of their orbitals and their ability to display a 3+ oxidation state [2, 3]. The results of researchers show that rare-earth material properties changes with doping or co-doping of them with oxides because of their crystalline structure variation.

Rare-earth-based materials were generally applied in high and low temperature fuel cells [4], gas storage/separation, chemical sensing [5], heterogeneous photocatalysis [6], wastewater treatment [7], fluorescence [8], photoelectro-chemical Water [9], etc. Nanostructure of rare-earth material is appropriate for these applications that can be synthesized with different methods for example coprecipitation, hydrothermal, sol-gel, combustion, stearic acid route, pechini, cathode plasma, electrolysis, co-ions complexation, molten salt and other approaches [7]. Each of these methods has its own advantages and disadvantages which are discussed in the following.

This study is being focused on heterogeneous photocatalytic activity of rare-earth-based material and their methods of preparation.

2. Modification of rare-earth-based materials

Efficiency of nanostructured compounds depend their grain size, morphology, structure and components. Study on properties of rare-earth oxide and doped with other material has increased due to their wide usages in various fields [10]. In order to develop properties, excessive effort has been done in the modification of morphology and structure of rare-earth materials.

Praseodymium oxide (Pr_6O_{11}), neodymium oxide (Nd_2O_3), holmium oxide (Ho_2O_3), $Dy_2Ce_2O_7$, $Nd_2Zr_2O_7$-Nd_2O_3, NdOCl-$Nd_2Sn_2O_7$-SnO_2, $Nd_2Sn_2O_7$-SnO_2, $Pr_2Ce_2O_7$, $Nd_2Sn_2O_7$, $Nd_2Zr_2O_7$-ZrO_2, $Nd_2Zr_2O_7$, $Dy_2Sn_2O_7$-SnO_2, Ho_2O_3-SiO_2, Nd_2O_3-SiO_2, are various families of rare-earth-based materials that were synthesized by researchers [10–20] to modify their properties and other modified rare-earth materials will be presented in the following.

2.1 Rare-earth-based coordination polymers

Rare-earth-based coordination polymers (CPs) have been studied due to their expanded structures and applications [5]. These materials contain catalytic active sites in both the organic linkers and metal centers. The modified of new and selective heterogeneous catalysts is suitable for photocatalytic processes, mainly in the oxidation of organic compounds.

Aguirre-Díaz et al. [5] synthesized five new lanthanide CPs (Ln-CPs) by solvothermal method using La, Nd, Sm, Ho, and Er as metal sources in order to offer a remarkable platform which contains antennas and catalytic active centers to achieve solar-energy conversion as green alternatives.

Also Le Natur et al. [21] synthesized a series of hexanuclear complexes based coordination polymers (CPs). They showed that CPs contain rare-earth material exhibit very high optical, luminescent properties and tunable systems at low cost. Therefore they demonstrated that this modified materials had high photocatalytic activity.

2.2 Rare-earth-based metal organic frameworks

Metal organic frameworks (MOFs) are crystalline porous materials with bridging organic ligands, which inorganic metal place in center. Nevertheless, applications of MOFs were limited because of their weak coordination capability and the stability. To solve this problem, Feng et al. [22] synthesized stable and functional 8-connected hexanuclear rare-earth MOFs platform based on $Re_6O_4(OH)_4(COO)_8$ Clusters. They showed that these structural characteristics cause that MOFs perform as a multipurpose platform for future applications, including recognition and high adsorption.

3. Methods for preparing rare-earth-based nanostructures

Shape of material as well as dimension will be influenced by the preparation method of nanoscale compounds. Therefor selection of technique is essential to control the shape of nanostructure [7]. In this section, we study some of methods for preparing rare-earth-based nanostructures.

3.1 Solid state reaction

In the solid state reaction as conventional method for synthesis, rare-earth oxide and other material are mixed using mechanical instrument and the achieved powders are calcined at very high temperature for very long time [23].

Eu^{3+} doped $Gd_2Mo_3O_9$ was prepared using solid-state reaction method and Na_2CO_3 as flux by Xiaoxia [24]. The results showed that the luminescent and photocatalytic properties depend to flux content and sintering temperature.

3.2 Hydrothermal

In the hydrothermal method, materials solutions are placed in the autoclave under high temperature and pressure conditions therefore the hydroxides are precipitated and the powders are formed through nucleation and growth. This method is suitable, simple, and effective controlled synthetic process and cheap because water is solvent [25].

Li et al. [26] used the low temperature hydrothermal method to synthesize nanosized $Y_2Sn_2O_7$ composite with the particle size from 10 to 300 nm using different organic agents. They demonstrated that these nanocomposites can be shown photocatalytic properties in water purification.

3.3 Combustion

When metal precursors (oxidants) and fuel react using combustion, these reactions can cause to produce rare-earth-based powders. In this method, a mixed solution consists of rare-earth material as precursors and urea as fuel were prepared then other solution (pH control agent) was added to this solution for gel formation. The prepared gel was dried and heated at high temperature to obtain rare-earth-based nanostructures [7, 27].

Zinatloo et al. [13] used combustion method to fabricate $Nd_2Sn_2O_7$-SnO_2 nanocomposites as novel and non-toxic biofuel using extract of pineapple. They showed that produced nanocomposites had high photocatalytic activity to degrade environmental pollution.

3.4 Coprecipitation

In the coprecipitation technique, the hydroxides are instantaneously precipitated from precursor solutions via adding a base solution. The formed hydroxide precipitates is then calcined to prepare rare-earth nanostructures. Ziarati et al. [28] synthesized $NiFe_{2-x}Eu_xO_4$ nanostructure using coprecipitation method and calcination. They showed that samples had high yields in presence of the rare-earth-based catalyst.

3.5 Sol-gel

In the sol-gel technique, hydrolysis and poly-condensation reactions of metal alkoxides occur to form gels. Afterwards, the achieved gels are calcined to obtain rare-earth nanoparticles. The sol-gel technique is low-cost, simple, friendly to the environment and less complicated than the others [7].

Li et al. [2] synthesized $LnFeO_3$ (Ln = Pr, Y) powders using sol-gel method with utilizing, lanthanide metal (Ln = Pr, Y) with the assistance of glycol at 800°C for 4 h. Photodecomposition rate of RhB was improved by $LnFeO_3$(Ln = Pr, Y) nanoparticles.

Nowadays, there are different ways for the production of rare-earth nanostructure such as sol-gel combined electrospinning, complex precipitation, pechini polymeric route, molten salt synthesis, co-ions complexation method, Cathode plasma electrolysis, Complex precipitation method, Precursor route and Floating zone technique.

3.6 Advantages and disadvantages of these methods

In this section, the advantages and disadvantages of each preparation technique have been discussed, briefly. Some nanostructured samples in this chapter have been synthesized using various methods by our research group.

There are different wet chemistry techniques to synthesize rare-earth-based nanostructures with suitable properties. In the wet chemistry technique, materials are mixed with together therefore the final powders have excellent compositional homogeneity. However, the solid state reaction technique is one of them that can cause the undesirable properties of final powders such as poor control of morphology and particle size and nonhomogeneity of distribution [23].

Hydrothermal method is homogeneous nucleation processes as remarkable advantage that can cause to synthesize particles with small size without the calcination stage. However, reaction rate and formation of rare-earth nanostructures are slow thus, a moderate heating stage at low temperature is necessary after the hydrothermal stage to accelerate the solid reaction [25].

The combustion technique can cause to synthesize powders with high purity, small size and uniform distribution. Nevertheless, the combustion reaction is not easy to control. Also, the combustion reaction can lead to agglomeration of the formed nanoparticles due to high temperature [27].

Coprecipitation method can lead to synthesize high purity compounds with a homogeneous composition at relatively low temperature but the obtained powders have large size. Thus a grinding stage is essential to synthesize a compound with small size [29, 30].

The sol-gel method is as an appropriate technique for compositional controlling the metallic species that the final powders are prepared through calcining stage with small size [31].

The type of method is effective on the characteristics of the powder therefore, selection of the right method is essential to prepare the powder with small particles, high uniform and purity. Also energy and cost is important to choose method.

4. Application of rare-earth-based material

Rare-earth-based materials were used in different fields. Some of rare-earth materials have been used for multipurpose applications such as gas sensors, solar cells, rechargeable batteries, supercapacitors, and transparent conducting electrodes, electronic and optical properties. The rare earth doped with some semiconductors such as SnO_2 can be used for temperature sensing. Recently, Eu^{3+}-ions doped SnO_2 has attracted the research attention as a candidate for thermometry applications [5].

Also, rare-earth-based perovskite oxides can be applied as catalysts for low-temperature fuel cells. The dimensions of the unit cell can be changed by varying the lanthanide ion which this unique property is an interesting characteristic of rare-earth perovskites ($LnMO_3$) [2].

According to the information of other researchers, rare-earth-based perovskites have been applied as fuel cell catalysts for methanol oxidation and oxygen reduction in alkaline medium. The results showed that these materials enhanced the kinetics of the methanol oxidation reaction in alkaline medium. On the other hand rare-earth-based perovskite oxides were used as cathode and anode catalysts for oxygen reduction in alkaline medium [9].

DOI: http://dx.doi.org/10.5772/intechopen.85525

Rare-earth-based materials such as CeO_2 also were added as an additive for Nafion membranes to improve their chemical stability. Hydrogen peroxide and its decomposition produce $HO^{\cdot}$ and $HOO^{\cdot}$ during the fuel cell operation that they have been considered as a key factors leading to membrane degradation therefore, CeO_2 nanoparticles act as mitigating free-radical-induced or free radical scavenger in biological systems due to their higher surface area and ability to undergo faster redox reaction. All studies suggested that CeO_2 nanoparticles have incredible potential to improve membrane permanence [4].

In this section, we study a brief review of rare-earth-doped materials, rare-earth-based oxide-oxide composites, metal-modified semiconductors and mixed-oxide materials in the fields of photocatalytic application as solar energy generation (**Figure 1**) [9].

Rare-earth-based material play a key role in important photocatalytic and energy production processes such as catalysts. Rare-earth material such as the lanthanide elements can perform as catalysts to enhance the efficiency by:

1. Developing properties of the catalyst surface,

2. Improving the thermal stability of catalytic oxides,

3. Enhancing the catalytic efficiency due to their redox capabilities and conductivity development,

4. Increasing oxygen uptake [9].

After a brief introduction of rare-earth materials and their properties ions and oxide materials. In the next sections will be discussed perspectives of rare-earth materials in photocatalytic processes and their potential as active catalysts or support materials.

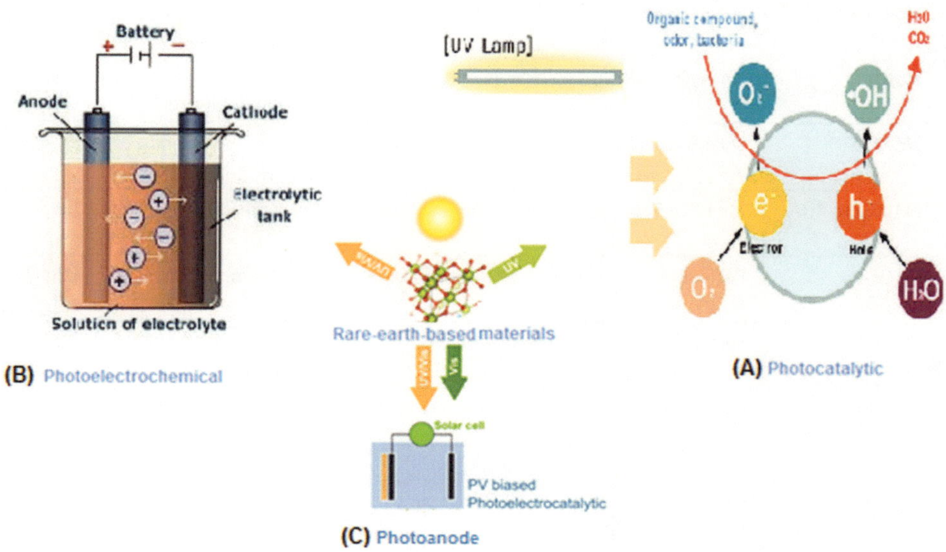

Figure 1.
Applications of rare-earth-based materials in solar energy generation: (A) photocatalytic, (B) photoelectrochemical, (C) cell approach (photoelectrocatalytic).

4.1 Heterogeneous photocatalysis

One of the progressive knowledge is heterogeneous photocatalysis that can be applied in diverse fields. Photocatalytic materials are used to degrade organic and inorganic pollutants in both the vapor and liquid phases [6].

When a photon is absorbed by rare-earth-based material using the energy gap of the materials, electron-hole pairs are generated in the photocatalysis mechanism.

Photocatalyst	Synthetic method	Light source	Removal percentage (%)	Ref.
RVO_4; R 5 La, Ce, Nd, Sm, Eu and Gd	Sonication	8 W mercury UV lamps of 365 nm wavelength	70% of methylene blue after 90 min illumination	[1]
$LnFeO_3$ (Ln = Pr, Y)	Sol-gel	500 W Xe lamp	96% of RhB after 2 h illumination	[2]
La-doped TiO_2	Sol-gel	UV irradiation at specific wavelength, 254 or 365 nm	≥90% of methylene blue after 90 min illumination	[8]
$Dy_2Ce_2O_7$	Green synthesis using juice of Punica	125 W Osram lamp	92.8% of methylene blue after 70 min illumination	[10]
$Dy_2Sn_2O_7$-SnO_2	Pechini	400 W mercury lamps	81% of methyl orange after 50 min illumination	[11]
$Nd_2Sn_2O_7$	Green synthesis using pomegranate juice	125 W Osram lamp	96.7% of methyl violet after 70 min illumination	[12]
$Nd_2Sn_2O_7$-SnO_2	Combustion	125 W Osram lamp	88.7% of rhodamine B after 70 min illumination	[13]
Nd_2O_3	Simple precipitation	125 W mercury lamps	89.6% of methylene blue after 90 min illumination	[14]
Ho_2O_3-SiO_2	Sonochemical	125 W mercury lamp	94.9% of methylene blue after 70 min illumination	[15]
Ho_2O_3	Sonochemical	125 W mercury lamp	82.2% of erythrosine after 70 min illumination	[16]
Nd_2O_3-SiO_2	Sonochemical	125 W Osram lamp	93.1% of methyl violet after 70 min illumination	[17]
$Nd_2Zr_2O_7$–ZrO_2	Modified Pechini	400 W mercury lamps	85% of methylene blue after 50 min illumination	[20]
$Pr_2Ce_2O_7$	Pechini	400 W mercury lamps	92.1% of methyl orange after 50 min illumination	[35]
$GdFeO_3$	Sol-gel	500 W Xe lamp	100% of RhB after 2 h illumination	[36]
$Dy_2Ce_2O_7$	Green synthesis using *Ananas comosus*	125 W Osram lamp	91.7% of black T after 1 h. illumination	[37]
$Dy_2Ce_2O_7$	Green synthesis using *Vitis vinifera* juice	125 W Osram lamp	92.4% of methyl orange after 80 min illumination	[38]
Nd_2O_3	Hydrothermal	UV light from the 400 W mercury lamps	79% of black T after 100 min illumination	[39]

Table 1.
Rare-earth-based materials for photocatalytic application in the degradation of organic pollutants.

Therefore light absorption with energy equal to or greater than the band gap of the material can cause an electron to be excited from the valence band to the conduction band [32]. These electrons and holes can be recombination with together that it is a major limiting factor to control its photocatalytic efficiency in the degradation of pollutant. Therefore, selection of material is a major challenge in heterogeneous photocatalysis that rare-earth-based nanostructure can be increased the efficiency of the photocatalysts [33]. To improve the efficiency of photocatalysis and reduce the recombination of electrons and holes, modification of rare-earth-based material is necessary. Therefore these materials have been doped with various transition metal ions [34]. The photocatalytic mechanism to degrade the pollutant contaminant may write as below [35]:

$\text{Prepared nanostructures} + h\upsilon \rightarrow \text{Prepared nanostructures} \ (e^-_{(CB)} + h^+_{(VB)})$.
$OH^- + h^+_{(VB)} \rightarrow OH^{\cdot}$
$O_2 + e^-_{(CB)} \rightarrow O_2^{\cdot -}$
$O_2^{\cdot -} + H^+ \rightarrow HO_2^{\cdot}$
$HO_2^{\cdot} + HO_2^{\cdot} \rightarrow O_2 + H_2O_2$
$H_2O_2 + O_2^{\cdot -} \rightarrow OH^{\cdot -} + OH^{\cdot} + O_2$
$OH^{\cdot} + \text{Organic} \rightarrow \text{Intermediates} \rightarrow CO_2 + H_2O + \text{Products}$

4.2 Photocatalytic applications

Rare-earth-based materials are applied as catalysts in different industrial applications, e.g., petroleum refining (lanthanum chloride), automobile emission reduction (cerium oxide), organic polymerization (neodymium versatate), hydrogen and synthesis gas production (ceria, lanthania), plastics decomposition, and dechlorination (samarium iodide), solar thermochemical water splitting and photocatalytic reactions have been investigated thoroughly, especially for La^-, Ce^-, Eu^-, and Gd^- modified materials [9].

The mixed oxide photocatalysts have attracted attention because of their abilities to remove pollutant or split water into H_2 and O_2 under solar irradiation. Among the mixed oxides, rare-earth-based materials contained mixed oxide photocatalysts are important, because the 4f-levels of lanthanide elements play a key role in photocatalytic reactions. Some researches are shown in **Table 1** that they include photocatalysts of rare-earth elements for degradation of organic pollutants. For example, Zinatloo et al. [13] synthesized $Nd_2Sn_2O_7$-SnO_2 nanocomposites as high-efficiency visible-light responsive photocatalyst through an environment-friendly procedure (Green synthesis) using extract of pineapple. Extract of pineapple that employed in this research, is natural source of sugar (glucose as well as fructose) and harmless for the environment.

5. Conclusions

This book chapter includes the recent advances related to synthesis method and photocatalytic applications of rare-earth-based nanostructures. Several techniques for preparation of rare-earth-based nanostructures have been introduced to achieve nanostructure with desirable characterizes.

Author details

Sahar Zinatloo-Ajabshir[1]* and Zahra Sayyar[2]

1 Department of Chemical Engineering, University of Bonab, Bonab, Iran

2 Faculty of Chemical Engineering, Sahand University of Technology, Tabriz, Iran

*Address all correspondence to: s.zinatloo@bonabu.ac.ir

References

[1] Selvan RK, Gedanken A, Anilkumar P, Manikandan G, Karunakaran C. Synthesis and characterization of rare earth orthovanadate (RVO_4; Rá = La, Ce, Nd, Sm, Eu & Gd) nanorods/nanocrystals/nanospindles by a facile sonochemical method and their catalytic properties. Journal of Cluster Science. 2009;**20**:291-305

[2] Li L, Zhang M, Tian P, Gu W, Wang X. Synergistic photocatalytic activity of $LnFeO_3$ (Ln = Pr, Y) perovskites under visible-light illumination. Ceramics International. 2014;**40**:13813-13817

[3] Bingham S, Daoud WA. Recent advances in making nano-sized TiO_2 visible-light active through rare-earth metal doping. Journal of Materials Chemistry. 2011;**21**:2041-2050

[4] Antolini E, Perez J. The use of rare earth-based materials in low-temperature fuel cells. International Journal of Hydrogen Energy. 2011;**36**: 15752-15765

[5] Aguirre-Díaz LM, Snejko N, Iglesias M, Sánchez F, Gutiérrez-Puebla E, Monge MÁ. Efficient rare-earth-based coordination polymers as green photocatalysts for the synthesis of imines at room temperature. Inorganic Chemistry. 2018;**57**(12):6883-6892

[6] Marcì G, García-López EI, Palmisano L. Heteropolyacid-based materials as heterogeneous photocatalysts. European Journal of Inorganic Chemistry. 2014; **2014**:21-35

[7] Zinatloo-Ajabshir S, Salavati-Niasari M, Sobhani A, Zinatloo-Ajabshir Z. Rare earth zirconate nanostructures: Recent development on preparation and photocatalytic applications. Journal of Alloys and Compounds. 2018;**767**: 1164-1185

[8] Anandan S, Ikuma Y, Murugesan V. Highly active rare-earth-metal La-doped photocatalysts: Fabrication, characterization, and their photocatalytic activity. International Journal of Photoenergy. 2012;**2012**: 921412

[9] Leduc J, Gönüllü Y, Raauf A, Fischer T, Mathur S. Rare-earth-containing materials for photoelectrochemical water splitting applications. In: Semiconductors and Semimetals. Vol. 97. United States: Elsevier; 2017. pp. 185-219

[10] Zinatloo-Ajabshir S, Salehi Z, Salavati-Niasari M. Green synthesis of $Dy_2Ce_2O_7$ ceramic nanostructures using juice of Punica granatum and their efficient application as photocatalytic degradation of organic contaminants under visible light. Ceramics International. 2018;**44**:3873-3883

[11] Zinatloo-Ajabshir S, Morassaei MS, Salavati-Niasari M. Facile fabrication of $Dy_2Sn_2O_7$-SnO_2 nanocomposites as an effective photocatalyst for degradation and removal of organic contaminants. Journal of Colloid and Interface Science. 2017;**497**:298-308

[12] Zinatloo-Ajabshir S, Morassaei MS, Salavati-Niasari M. Nd2Sn2O7 nanostructures as highly efficient visible light photocatalyst: Green synthesis using pomegranate juice and characterization. Journal of Cleaner Production. 2018;**198**:11-18

[13] Zinatloo-Ajabshir S, Morassaei MS, Salavati-Niasari M. Facile synthesis of $Nd_2Sn_2O_7$-SnO_2 nanostructures by novel and environment-friendly approach for the photodegradation and removal of organic pollutants in water. Journal of Environmental Management. 2019;**233**: 107-119

[14] Zinatloo-Ajabshir S, Mortazavi-Derazkola S, Salavati-Niasari M. Nd_2O_3 nanostructures: Simple synthesis, characterization and its photocatalytic

degradation of methylene blue. Journal of Molecular Liquids. 2017;**234**:430-436

[15] Zinatloo-Ajabshir S, Mortazavi-Derazkola S, Salavati-Niasari M. Simple sonochemical synthesis of Ho_2O_3-SiO_2 nanocomposites as an effective photocatalyst for degradation and removal of organic contaminant. Ultrasonics Sonochemistry. 2017;**39**: 452-460

[16] Zinatloo-Ajabshir S, Mortazavi-Derazkola S, Salavati-Niasari M. Sono-synthesis and characterization of Ho_2O_3 nanostructures via a new precipitation way for photocatalytic degradation improvement of erythrosine. International Journal of Hydrogen Energy. 2017;**42**:15178-15188

[17] Zinatloo-Ajabshir S, Mortazavi-Derazkola S, Salavati-Niasari M. Nd_2O_3-SiO_2 nanocomposites: A simple sonochemical preparation, characterization and photocatalytic activity. Ultrasonics Sonochemistry. 2018;**42**:171-182

[18] Zinatloo-Ajabshir S, Salavati-Niasari M. Nanocrystalline Pr_6O_{11}: Synthesis, characterization, optical and photocatalytic properties. New Journal of Chemistry. 2015;**39**:3948-3955

[19] Zinatloo-Ajabshir S, Salavati-Niasari M, Zinatloo-Ajabshir Z. $Nd_2Zr_2O_7$-Nd_2O_3 nanocomposites: New facile synthesis, characterization and investigation of photocatalytic behaviour. Materials Letters. 2016; **180**:27-30

[20] Zinatloo-Ajabshir S, Zinatloo-Ajabshir Z, Salavati-Niasari M, Bagheri S, Hamid SBA. Facile preparation of $Nd_2Zr_2O_7$–ZrO_2 nanocomposites as an effective photocatalyst via a new route. Journal of Energy Chemistry. 2017;**26**: 315-323

[21] Le Natur F, Calvez G, Daiguebonne C, Guillou O, Bernot K, Ledoux J, et al. Coordination polymers based on heterohexanuclear rare earth complexes: Toward independent luminescence brightness and color tuning. Inorganic Chemistry. 2013;**52**:6720-6730

[22] Feng Y, Xin X, Zhang Y, Guo B, Li F, Kong X, et al. Rational design and synthesis of hexa-nuclear rare earth the-a MOFs platform based on $Re_6O_4(OH)_4(COO)_8$ clusters. Crystal Growth & Design. 2019;**19**(3):1509-1513

[23] Tong Y, Zhu J, Lu L, Wang X, Yang X. Preparation and characterization of $Ln_2Zr_2O_7$ (Ln = La and Nd) nanocrystals and their photocatalytic properties. Journal of Alloys and Compounds. 2008; **465**:280-284

[24] Xiaoxia Z, Xiaojun W, Baojiu C, Qingyu M, Weihua D, Guozhong R, et al. Photoluminescence characteristics of $Gd_2Mo_3O_9$: Eu phosphor particles by solid state reaction method. Journal of Rare Earths. 2007;**25**:15-18

[25] Byrappa K, Adschiri T. Hydrothermal technology for nanotechnology. Progress in Crystal Growth and Characterization of Materials. 2007;**53**:117-166

[26] Li K, Wang H, Yan H. Hydrothermal preparation and photocatalytic properties of Y2Sn2O7 nanocrystals. Journal of Molecular Catalysis A: Chemical. 2006;**249**:65-70

[27] Tong Y, Zhao S, Feng W, Ma L. A study of Eu-doped $La_2Zr_2O_7$ nanocrystals prepared by salt-assistant combustion synthesis. Journal of Alloys and Compounds. 2013;**550**:268-272

[28] Ziarati A, Sobhani-Nasab A, Rahimi-Nasrabadi M, Ganjali MR, Badiei A. Sonication method synergism with rare earth based nanocatalyst: Preparation of $NiFe_2$–$xEuxO_4$ nanostructures and its catalytic applications for the synthesis of benzimidazoles, benzoxazoles, and

benzothiazoles under ultrasonic irradiation. Journal of Rare Earths. 2017; **35**:374-381

[29] Duarte W, Meguekam A, Colas M, Vardelle M, Rossignol S. Effects of the counter-cation nature and preparation method on the structure of $La_2Zr_2O_7$. Journal of Materials Science. 2015;**50**: 463-475

[30] Kong L, Karatchevtseva I, Gregg DJ, Blackford MG, Holmes R, Triani G. A novel chemical route to prepare $La_2Zr_2O_7$ pyrochlore. Journal of the American Ceramic Society. 2013;**96**: 935-941

[31] Mahdiani M, Sobhani A, Salavati-Niasari M. Enhancement of magnetic, electrochemical and photocatalytic properties of lead hexaferrites with coating graphene and CNT: Sol-gel auto-combustion synthesis by valine. Separation and Purification Technology. 2017;**185**:140-148

[32] Zou Z, Ye J, Arakawa H. Role of R in Bi_2RNbO_7 (R = Y, rare earth): Effect on band structure and photocatalytic properties. The Journal of Physical Chemistry B. 2002;**106**:517-520

[33] El-Bahy ZM, Ismail AA, Mohamed RM. Enhancement of titania by doping rare earth for photodegradation of organic dye (direct blue). Journal of Hazardous Materials. 2009;**166**: 138-143

[34] Liang C-H, Li F-B, Liu C-S, Lü J-L, Wang X-G. The enhancement of adsorption and photocatalytic activity of rare earth ions doped TiO2 for the degradation of Orange I. Dyes and Pigments. 2008;**76**:477-484

[35] Zinatloo-Ajabshir S, Salehi Z, Salavati-Niasari M. Preparation, characterization and photocatalytic properties of $Pr_2Ce_2O_7$ nanostructures via a facile procedure. RSC Advances. 2016;**6**:107785-107792

[36] Li L, Wang X, Lan Y, Gu W, Zhang S. Synthesis, photocatalytic and electrocatalytic activities of wormlike $GdFeO_3$ nanoparticles by a glycol-assisted sol-gel process. Industrial and Engineering Chemistry Research. 2013; **52**:9130-9136

[37] Zinatloo-Ajabshir S, Salehi Z, Salavati-Niasari M. Green synthesis and characterization of $Dy_2Ce_2O_7$ nanostructures using Ananas comosus with high visible-light photocatalytic activity of organic contaminants. Journal of Alloys and Compounds. 2018; **763**:314-321

[38] Zinatloo-Ajabshir S, Salehi Z, Salavati-Niasari M. Green synthesis and characterization of $Dy_2Ce_2O_7$ ceramic nanostructures with good photocatalytic properties under visible light for removal of organic dyes in water. Journal of Cleaner Production. 2018;**192**: 678-687

[39] Zinatloo-Ajabshir S, Mortazavi-Derazkola S, Salavati-Niasari M. Schiff-base hydrothermal synthesis and characterization of Nd_2O_3 nanostructures for effective photocatalytic degradation of eriochrome black T dye as water contaminant. Journal of Materials Science: Materials in Electronics. 2017; **28**:17849-17859

Chapter 6

Radiative Transference Equation Algorithm as an ANSYS® User-Defined Function for Solar Technology Applications

Diana Barraza-Jiménez, Adolfo Ruiz-Soto, Sandra Iliana Torres-Herrera, Elva Marcela Coria-Quiñones, Raúl Armando Olvera-Corral, David José Romero-Soto and Manuel Alberto Flores-Hidalgo

Abstract

Heat effects in photocatalytic reactor applications are discussed and a case study is analyzed where sunlight is used to activate a chemical reaction to degrade water pollutants. Heat is produced in the light-capturing process, and heat effects need to be better understood during the device design process. Radiative transfer equation (RTE) is the guiding equation used to calculate radiation proliferation in participating media, and it is used to describe the balance of radiative energy transport in the participating media including the interactions caused by different processes such as absorption, scattering, and emission, which also are subject to additional phenomena like weakening and magnification. This equation plays an important role in the design process since it may be included in the simulation process to represent the sunlight heat effects in the different photocatalytic reactor components. In this chapter, it is explained how to build a simplified algorithm to incorporate the RTE in a numerical calculation during the design of a photocatalytic reactor using the commercial software ANSYS®. In addition, simplifications are explained that enable the program to coordinate some coefficients such as absorption and dispersion so their effects are included within the numerical calculation. A user-defined function is presented in the end of the chapter as a usable algorithm in ANSYS® program with acceptable results for photocatalytic reactors.

Keywords: radiative transfer equation, light source energy, radiation energy, ANSYS®, photocatalysis

1. Introduction

Solar technology includes a wide variety of developments in environmental applications that include photovoltaic cells and photocatalytic devices, among others. Sunlight produces heat that affects the devices, components, and different processes that are carried on. The main interest in this proposal is approach heat effects in photocatalytic reactor applications where sunlight is used to activate a chemical reaction to degrade water pollutants. Heat is produced in the light-capturing process, and heat effects need to be better understood during the device design process [1, 2].

Radiative transfer equation (RTE) is the guiding equation used to calculate radiation proliferation in participating media, and it is used to describe the balance of radiative energy transport in the participating media including the interactions caused by different processes such as absorption, scattering, and emission which also are subject to additional phenomena like weakening and magnification [3, 4]. This equation plays an important role in the design process since it may be included in the simulation process to represent the sunlight heat effects in the different photocatalytic reactor components. In this chapter, it is explained how to build a simplified algorithm to incorporate the RTE in a numerical calculation during the design of a photocatalytic reactor using the commercial software ANSYS Fluent® [5]. This represents a numerical solution of one of the possible presentations of the RTE that can be easily programmed and incorporated in extensive calculations. In addition, simplifications are explained which enable the program to coordinate some coefficients such as absorption and dispersion so their effects are included within the numerical calculation. A user-defined function is presented in this chapter as a usable algorithm in ANSYS® program with acceptable results for photocatalytic reactors.

Simplifications have been done in order to focus on reaching the main goal of this chapter which is study the requirements to have a programmed UDF that resolves the radiation transfer energy equation. It is difficult to find out a generalized program so this chapter will assume the mathematical model applies for a photocatalytic reactor with a light source so the light energy is transferred into the system of study. There might be a transparent wall that the light needs to go through as well as water or polluted water as the fluid media. Above all, the goal is simplifying as much as possible the mathematical model and developments within the chapter in order to focus in an analysis on how to program RTE equation.

The program will be written similarly compared to C language, but the language is considered a property of ANSYS® but more specifically in ANSYS Fluent®. That means the programming language is its own language for ANSYS Fluent®, but it follows a syntaxis and structure very similar to C language. It is highly recommended to have C language programming skills and ANSYS Fluent® user-defined functions training. These skills may be achieved after one is an advanced user in ANSYS Fluent® so the programming can be coupled with CFD calculations and then the problem at hand with hydrodynamics numerical results will be complemented with light source effects, which at the end is the main purpose of this chapter.

2. Radiative energy basic concept

As we shall see, radiative heat transfer rates are generally proportional to differences in temperature to the fourth (or higher) power, i.e.,

$$q \propto T^4 - T_\infty^4. \tag{1}$$

Then, if temperature rises, radiative heat transfer relevance grows and may become the dominant factor over conduction and convection phenomena at very high temperatures. In this way, one may understand thermal radiation is a meaningful factor in combustion applications (fires, furnaces, rocket nozzles, engines, and others) and in nuclear reactions (such as in a fusion reactor or in nuclear bombs), during atmospheric reentry of space vehicles, among others. Since contemporary technology's main goal is an increased efficiency, it is expected temperature will increase accordingly, and this situation will reflect in making thermal radiation even more important. In addition, emergent applications gaining importance are solar energy and greenhouse effect (both as a result of our sun's heating effects). Derived from the previous effects, one of the human kind problems in these days is global warming which is provoked by solar energy absorption mainly by carbon dioxide released into our atmosphere primarily by fuel burning technological devices created by men.

In the same way that factors such as vacuum and high temperature become more important for thermal radiation effects, also the analysis methodology to study thermal radiation becomes more complicated or at least requires to involve additional aspects if compared to analyses in normal conditions. One can say, in general, thermal radiation is a comprehensive matter to analyze. The average distance traveled by a photon previously to interact with a molecule is also known as the photon mean free path may be as short as 10^{-10} m (absorption process) but may be very long in the possible maximum distance reaching up to 10^{+10} m or even more (e.g., sunlight over the Earth's surface). Then, it is natural to conclude conservation of energy will not be applicable on an infinitesimal volume, but must be applied on the entire volume of interest. Therefore, an integral equation with seven independent variables (radiation frequency, three space coordinates, two coordinates to define photons travelling direction, and time) may be proposed.

The effects caused by the radiative properties of materials complicate even more the thermal radiation analysis. Measuring the radiative properties is usually difficult and displays erratic behavior most of the times. Properties for liquids and solids depend on a really thin surface layer, so it is natural that these properties may be affected if there is a surface preparation for this purpose and then the properties may change when this occurs. When it refers to gases, all radiative properties may vary by much with wavelength if another dimension is added to the governing equation. It may be assumed as well that this equation is nonlinear and rarely can be considered otherwise. There are different methods to solve the governing energy equation for radiative transport but escape the scope of this chapter, and the reader is referred to related literature such as whole books that deal with the mathematical details of this equation.

3. Radiative energy transfer equation

The radiative equation has different applications, for example, in aerospace, physics, and astrophysics. Fundamentally, this equation describes a field of radiation called spectral radiance in radiometric terms.

For example, to determine the radiative flux onto a surface (or at any point inside the medium) requires knowledge of the radiative intensity at that point, for all directions and for all wavelengths or wavenumbers. In an application such as a photocatalytic reactor, the importance of energy relates to the intensity field inside a participating medium that is governed by the radiative transfer equation, which is a radiative energy balance along a thin group of rays, but for a medium that emits, absorbs, and/or scatters radiation.

The equation of radiative transfer describes mathematically interactions derived from the propagation of radiation through a medium that means that phenomena such as absorption, emission, and scattering processes are contained in an equation.

The radiative transfer equation (RTE) describes the radiation intensity I_λ at any position along a ray path through a medium. This equation can be applied to evaluate the radiation field in the heterogeneous system constituted by a photocatalyst particulate suspension in water. RTE equation may be presented in different forms; in our case we are interested in the application for a photocatalytic reactor. In such application the energy of a light source is used to activate a photocatalyst that will participate in the chemical degradation of a pollutant in water. In a reactor like this one, the energy path starts at the light source, passes through a transparent wall that may be a glass or quartz or similar, and then travels through polluted water until the energy reaches the photocatalyst. Then, for a participating medium, with absorption and scattering (no emission is considered), the RTE can be written as

$$\frac{dI_{\lambda,\Omega}}{ds} = \underbrace{-k_y I_{\lambda,\Omega}}_{\text{Absorption}} \underbrace{- \sigma_\lambda I_{\lambda,\Omega}}_{\text{Outgoing dispersion}} + \underbrace{\frac{\sigma_\lambda}{4\pi}\int_{\Omega'=4\pi} p(\Omega' \rightarrow \Omega) I_{\lambda,\Omega'} d\Omega'}_{\text{Incoming dispersion}} \quad (2)$$

where $I_{\lambda,\Omega}$ is the intensity of photons with wavelength λ, propagated in the direction Ω, k_λ is the volumetric absorption coefficient, $\sigma\lambda$ is the volumetric dispersion coefficient, and $p(\Omega' \rightarrow \Omega)$ is the phase function that describes the dispersed radiation direction distribution.

4. Numerical solution for radiative energy problems

Therefore, there are systems or devices where radiative heat transfer becomes more important with rising temperature levels and may be totally dominant over conduction and convection at very high temperatures. Resolution of RTE equation is complex for real systems such as a photochemical reactor. There are different approaches to resolve this equation from making assumptions of ideality for the system or using empirical data. Let's discuss how can a resolution be reached next.

4.1 An approach to solve RTE

Analytic solutions to the radiative transfer equation (RTE) exist for simple cases, but for more realistic media, with complex multiple scattering effects, numerical methods are required. Before reaching the numerical method approach, one may notice that in order to resolve Eq. (2), the volumetric scattering coefficient σ_λ and the phase function that describes the directional distribution of scattered radiation $p(\Omega' \rightarrow \Omega)$ are needed mainly due to the volumetric absorption coefficient κ_λ, which displays strong spectral dependence. From this calculation, the intensity of photons with wavelength λ propagated along direction Ω, and $I_{\lambda,\Omega}$ is the result that may be obtained. An integral on all the spatial directions may be used to account for the whole radiation field for any point of the reactor space, and by multiplying this value by the absorption coefficient, one may obtain the local volumetric rate of energy absorption (VREA).

4.2 Discrete ordinate method and numerical solution

Resolution of RTE may be reached using a numerical technique known as the discrete ordinate method (DOM). An advantage to use DOM is that this method is incorporated in ANSYS Fluent®. This method solves the RTE for a finite number of discrete solid angles, each one associated with a vector direction. For the case where

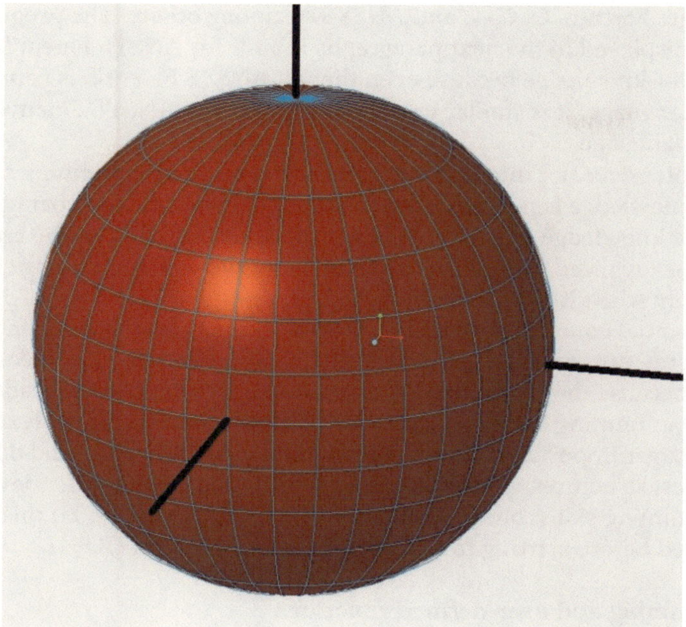

Figure 1.
Geometry assuming a spherical shape with an $n \times m$ control elements division of 10×10 for numerical calculations using DOM.

a computational tool such as ANSYS Fluent® is used, the spatial discretization of the computational region of interest may be taken from the mesh grid created during the discretization step required by this program. However, the directional discretization for the RTE has to be specified using an angular discretization of the sphere octant of $m \times n$ divisions (divisions should be enough to avoid the well-known "ray effect") and $p \times q$ pixelation (enough to overcome angle overhanging). ANSYS Fluent® has more models for radiation calculation incorporated that may be more simplified than DOM, but the discrete ordinate method is able to define optical processes considering wavelength; also, it considers dispersion effects, absorption, and emission of radiation in media and surfaces.

DOM may be able to resolve RTE based on a finite number of discrete solids with an angle and direction which may be related with a vector. ANSYS Fluent® provides this information by means of the meshing process which contains region location data needed by DOM, but direction of discretization for RTE evaluation should be specified by the user by means of an angular discretization. To address the direction of discretization, each octant in the 4π angular space may be discretized in $N_{\theta p} \times N_{\phi p}$ control angles. These angles relate with polar and azimuthal directions and are fundamental for RTE resolution. **Figure 1** displays an example of control elements.

Each control element may be subdivided $N_{\theta p} \times N_{\phi p}$ for a bigger discretization, and calculations will refer to the pixels generated either for incoming or outgoing energy. This is an optional technique to obtain a further division of the surface in case it is required.

4.3 Programming a numerical method to resolve RTE

Programming a method to resolve the RTE is a way to carry on with a numerical solution. There are different options to program the RTE; among the more common choices may be MATLAB which is a numerical computation system by itself with its own language, Fortran, C, C++, and ANSYS®, among others. The programming that will be displayed in the next paragraphs is built for ANSYS Fluent® so it needs some C or C++ knowledge because even though ANSYS Fluent® is considered to use its own language, it is similar to C or C++ given that ANSYS Fluent® itself was built in that language.

This chapter does not intent to educate the reader in programming but assumes some basic knowledge is needed in order to keep up with the next paragraphs. Also, there is some knowledge in ANSYS Fluent® needed; more likely, the reader might need to be an advanced user of this program since we will mention briefly some of the basics only when it is absolutely needed.

Some essential comments related to programming C language are included because it is relevant to understand UDFs in ANSYS Fluent®. These comments are really basic because they are intended to provide only a hint to the reader, but a C language programming book is highly recommended. So, in this context, this chapter will not cover important C programming topics such as while and do-while, unions, cycles, structures, and reading and writing files. For further developing your programming skills, one may consult a C book or guide [6, 7]; this is highly recommended before starting to build or program your own UDFs.

4.4 Programming and user-defined functions

The ANSYS Fluent® has a tool to input complementary code with customized functions or actions. This complementary code is known in ANSYS® user manual as user-defined functions (UDFs). User-defined functions are the more common tool to customize calculations in ANSYS Fluent®; however, programming a UDF is complicated, and it is considered an advanced skill for users of ANSYS Fluent®.

A UDF, in short, is a C or C++ small program that acts as a function so it can be easily loaded in the ANSYS Fluent® solver with the intention of enhancing its standard features. For example, a UDF may be used to:

- Specify boundary conditions, material properties, surface and volume reaction rates, source terms in ANSYS Fluent® transport equations, user-defined scalar (UDS) for source terms in transport equations, and diffusivity functions, among others.

- Once per iteration adjustment of specific computed values.

- Solution initialization.

- Execution of a UDF performs asynchronous calculations as required.

- End of an iteration function execution, computations after exiting from ANSYS Fluent®, or after loading a compiled library.

- Additional capabilities or computations during postprocessing.

- Additional capabilities or computations from existing ANSYS Fluent® models (additional to mathematical models already loaded like discrete phase model, multiphase mixture model, discrete ordinates radiation model).

For further knowledge in UDFs, the reader is referred to literature produced by ANSYS® or specialized books [8].

From the authors perspective, a UDF is a C program conformed by a set of macros developed specifically for ANSYS Fluent®; such macros may be considered small pieces of the program or built in functions specially for ANSYS Fluent®. If one takes a look at a UDF, a code will be able to identify a program in C or C++, but very few lines will be purely these languages because most lines are macros specifically developed for ANSYS Fluent®. Then, in conclusion, if one looks at the content of an UDF, he or she will find a few lines purely programmed in C plus a series of macros built in to work only in ANSYS Fluent® all of them ready to add up a specific computation complementary to ANSYS Fluent® capabilities.

4.5 Coupling the UDF to a multiphysics program suite

The UDF will provide the steps to numerically resolve RTE, but there is an important action that needs to be carefully performed, and that is the coupling of the UDF in the ANSYS Fluent®. The general theory in UDF development says a user-defined function may be programmed to execute steps in different stages of the problem numerical resolution. In this case, the need to resolve the RTE equation relies in the need to find out numerically the light energy influence in a photocatalytic reactor. To accomplish this purpose, the UDF will execute after hydrodynamic calculations have been completed.

Then, after the UDF is available, the user will have to perform a compilation of the program to make sure the syntaxis is correct. After compilation, the UDF is ready to be loaded into the Multiphysics program in this case ANSYS Fluent®. In fact, the compilation process is also performed within ANSYS Fluent® as well as the loading process where the UDF is coupled to the main code.

A last recommendation after the UDF is compiled and loaded into ANSYS Fluent® is to perform testing runs of the problem numerical resolution to make sure the UDF works well and is returning good numerical results.

5. User-defined functions to resolve the RTE

In this section, examples of UDFs that can be used to resolve numerically the RTE are presented.

5.1 UDF used to resolve the Henyey-Greenstein phase function

As mentioned in prior paragraphs, it is needed to resolve the phase function. Since the selected phase function was the Henyey-Greenstein, it is needed to develop a UDF to have this function resolved. Next, a proposed UDF to resolve this function in pretty basic terms is shown [9]. Additional comments have been added to explain further the purpose of the programmed line. In each line or command, or cycle, or macro, the user may add comments to help out understand the UDF design intent during possible reviews, corrections, tests, improvement, etc. In these examples, comments are added for didactic reasons but in a professional UDF if may be recommended much less comments and limited to explain the more relevant parts of the program. A solution for the selected phase function was the Henyey-Greenstein which is displayed in **Table 1**.

5.2 UDF to resolve radiation transfer from a light source

This UDF represents a numerical solution to the energy transfer from a light source to a reactor. The skeleton of this UDF was recycled from the Sandia National

```
/* Dear all: this is a UDF built specifically to resolve the
Henyey-Greenstein phase function for thermal radiation calculations*/ /* comments section*/
#include"udf.h" /* command to initiate a udf*/
DEFINE_SCAT_PHASE_FUNC(hgscat,c,f) /* DEFINE is a macro*/
{ /* Command to start a computation*/
real p = 0.; /* set up a data type*/
real g = 0.5161; /* set up a data type*/
p = (1-pow(g,2.0))/pow((1+g*g-2*g*c),1.5); /* Equation for calculation*/
return(p); /* output*/
} /* Command to end a computation */
```

Table 1.
Proposed UDF to resolve the Henyey-Greenstein phase function.

```
/* Tested UDF to integrate UV intensity over time along a particle trajectory. */
/* When volume fraction is very small and a cell is reached particles are deleted from the domain */
/*Low dosed particles are identified and residence times are obtained at the reactor outlet. */
/* comments section*/
#include "udf.h" /* command to initiate a udf*/
#include "dpm.h" /* command to call a udf*/
#include "sg_disco.h" /* command to access cell UV intensity values*/
#define C_DO(c,t)C_STORAGE_R_XV(c,t,SV_DO_IRRAD,0) /* DEFINE is a macro*/
#define fdout "doseout.txt" /* DEFINE is a macro*/
#define ftrack "trackfilters.txt" /* DEFINE is a macro*/
#define ftime "restime.txt" /* DEFINE is a macro*/
FILE *pdout; /* use file*/
FILE *ptrack; /* use file*/
FILE *ptime; /* use file*/
static real uv_intensity;
/* This macro integrates UV intensity over time along a particle trajectory */ /* comments section*/
/* This macro writes the UV dosage, residence time and ID for low dosed particles at reactor outlet. */ /*
comments section*/
DEFINE_DPM_OUTPUT(uv_output, header, fp, p, thread, plane) /* Define macro, variable set up*/
{
if (header) /* if-else loop*/
{
}
else
{
/* write the dosage values */ /* comments*/
pdout=fopen(fdout,"a"); /* delete previous files */ /* data type, comments*/
fprintf(pdout,"%10.6g \n",p->user[0]); /* output, comments*/
fclose(pdout); /* end computation*/
/* write the particle id of low dosed particles */ /* comments*/
ptrack=fopen(ftrack,"a"); /* delete previous files */ /* datatype, comments*/
if(p->user[0]<30.0) /* change this value accordingly */ /* if-else loop, comments*/
{
fprintf(ptrack,"%d,",p->part_id); /* output, calculations*/
}
fclose(ptrack); /* end computation*/
/* write the particle residence times */ /* comments*/
ptime=fopen(ftime,"a"); /* delete previous files */ /* data, calculation, comments*/
fprintf(ptime,"%10.6g \n",p->st ate.time); /* output, calculations*/
fclose(ptime); /* end computation*/
}
}
```

Table 2.
Proposed UDF to resolve the effects of a light source radiation energy to the reactor passing through transparent media, in our case, polluted water.

Laboratories website. It was updated to suit the open channel flow application and new features were added [10]. Additional comments have been added to explain further the purpose of the programmed line. **Table 2** displays a UDF to resolve the radiation energy as a result of the emission of a light source.

6. Conclusions

The mathematical model for numerical calculations of the effects of light into a photocatalytic reactor is discussed. Despite the difficulty of a generalized method, it is reached the goal on presenting a study of the requirements to have a programmed UDF that resolves the radiation transfer energy equation. Due to the difficulty to find out a generic code applicable in all situations, for this chapter, one may assume the mathematical model applies for a photocatalytic reactor with an external light source so the light energy is transferred into the system of study. There is a transparent wall that the light needs to go through as well as water or polluted water as the fluid media. For such system, the process to obtain a UDF that calculates numerically the effects of the light source in the reactor is explained. The program was written similarly compared to C language, but the language is considered a property of ANSYS® or more specifically in ANSYS Fluent®. That means the programming language is its own language for ANSYS Fluent®, but it follows a syntaxis and structure very similar to C language. It is highly recommended to develop C language programming skills and ANSYS Fluent® user-defined function training. These skills may be achieved after one is an advanced user in ANSYS Fluent® so the programming can be coupled with CFD calculations and then the problem at hand with hydrodynamics numerical results will be complemented with light source effects, which at the end is the main purpose of this chapter. The UDFs presented are displayed so the reader can understand its purpose and take advantage to develop their own program.

Acknowledgements

This work was financed by CONACyT (Mexican Science and Technology National Council) through 2015 CONACyT SEP-CB (Basic Science-Public Education Ministry) project fund 258553/CONACyT/CB-2015-01. Thanks go to the Scientific Computing Laboratory at FCQ-UJED for computational resources. Thanks go to the Academic Group UJED-CA-129 for valuable discussions.

Conflict of interest

The authors state that this research was completed without any conflicts of interest related with funding to develop the present work.

Author details

Diana Barraza-Jiménez[1], Adolfo Ruiz-Soto[1,3], Sandra Iliana Torres-Herrera[2], Elva Marcela Coria-Quiñones[1,3], Raúl Armando Olvera-Corral[1], David José Romero-Soto[1] and Manuel Alberto Flores-Hidalgo[1*]

1 Faculty of Chemistry Science, Juarez University of Durango State, Durango, Mexico

2 Faculty of Forestry Science, Juarez University of Durango State, Durango, Mexico

3 TecNM/Durango Institute of Technology, Durango, Mexico

*Address all correspondence to: maflores.hidalgo02@gmail.com

References

[1] Moreno J, Casado C, Marugán J. Improved discrete ordinate method for accurate simulation radiation transport using solar and LED light sources. Chemical Engineering Science. 2019; **205**:151-164

[2] Satuf ML, Brandi RJ, Cassano AE, Alfano OM. Experimental method to evaluate the optical properties of aqueous titanium dioxide suspensions. Industrial and Engineering Chemistry Research. 2005;**44**: 6643-6649

[3] Schneider J, Bahnemann D, Ye J, Puma GL, Dionysiou DD. Photocatalysis: Fundamentals and Perspectives. Cambridge, UK: Royal Society of Chemistry; 2016. pp. 351-366. ISBN: 978-1-78262-041-9

[4] Modest MF. Radiative Heat Transfer. 3d ed. Oxford, UK: Elsevier Inc; 2013. DOI: 10.1016/C2010-0-65874-3. ISBN: 978-0-12-386944-9

[5] ANSYS Fluent® Academic Research CFD; Release 19.1

[6] Kernighan B, Ritchie D. The C Programming Language. 2nd ed. New Jersey, USA: Prentice-Hall; 1988. ISBN: 0-13-110370-9

[7] Stephen GK. Practical Programming in C. 3rd ed. Indianapolis, USA: Sams Publishing; 2005. ISBN: 0-672-32666-3

[8] ANSYS® INC. ANSYS Fluent®. Customization Manual; 2017

[9] Casado C, Marugán J, Timmers R, Muñoz M, van Grieken R. Comprehensive multiphysics modeling of photocatalytic processes by computational fluid dynamics based on intrinsic kinetic parameters determined in a differential photoreactor. Chemical Engineering Journal. 2017;**310**(P2): 368-380

[10] Rajib Kumar S. Numerical simulation of an open channel ultraviolet waste-water disinfection reactor [thesis and dissertation repository]. Electronic; 2013. 1450 p. Available from: https://ir.lib.uwo.ca/etd/1450

Chapter 7

Effect of Annealing on Metal-Oxide Nanocluster

Naorem Khelchand Singh and Rajshree Rajkumari

Abstract

Recently, the development of optoelectronic devices based on metal-oxide nanocluster has attracted intensive research interest. Nanoclusters are suitable for these because of their large surface-to-volume ratio and the presence of abundant oxygen vacancies or trap states. Metal–oxides such as ZnO, In_2O_3, and TiO_2 synthesized using different technique produces high surface area films consisting of clusters and provides complete control over the film morphology. In this chapter, some of the metal oxides nanocluster film has investigated, and the effect of annealing on the structural, optical and electrical properties of the grown films when subjected to different annealing temperatures will be studied. Theoretically, these properties are presumed to improve after the heat treatment as the crystallinity, and the grain size of the film has increased due to the diminishing of oxygen vacancies. Thus, the greater surface-to-volume ratio, the better stoichiometry and higher level of crystallinity compared to bulk materials make nanocluster-based devices very promising for the mentioned application.

Keywords: nanomaterials, metal oxide semiconductor nanomaterials, fabrication technique, heat treatment, nano-devices

1. Introduction

Surface area and quantum confinement effects are the two important key factors that cause the properties of nanomaterials to differ significantly from the normal bulk materials. Among those nanomaterials, the metal oxide nanoparticles /nanoclusters exhibit unique physical and chemical properties due to their high surface area and nanoscale size. As the particle decreases, a greater amount of atoms are found at the surface compared to those inside. The particle size is influenced to the structural characteristics namely the lattice symmetry and cell parameter) as well as electronic properties in any material. Metal oxide nanocluster finds its promising applications in optical and sensors [1], optoelectronics [2], photocatalysts [3] and biomedical applications [4]. Till now, there have been a considerable number of reports for the fabrication of various metal oxide nanocluster, such as ZnO [5–8], TiO_2 [1, 3, 9], SnO_2 [10–13], In_2O_3 [14–16], CuO_2 [17–19], WO_3 [20, 21]. The reported technique for the synthesis includes (a) physical synthesis such as pulsed laser deposition [5], glancing angle deposition [2], cluster beam deposition [22], and others; (b) chemical synthesis such as sol-gel [9], hydrothermal decomposition [18], and others; (c) biological synthesis using agro wastes, algal media, plant extract, fungi, yeast, and others.

 IntechOpen

Metal oxide nanocluster in the context of high-performance optoelectronic devices based entirely on the structural, optical and electrical properties. The electrical conductivity of the device strongly depends on the surface defects, grain boundary, surface adsorbed oxygen and unsaturated dangling bonds. In fact, metal oxide nanocluster contains a considerable number of point and surface defects associated with the deviation from the stoichiometric composition. Moreover, the generation of oxygen vacancies in metal oxide nanocluster hampers the efficiency and accuracy of the device. So, in order to reduce the deformation in the surface structure and high defect density related to oxygen vacancies, heat treatment is necessary. Many studies reported on synthesizing metal oxide nanocluster followed by annealing in the different atmosphere. They have shown that annealing at an optimal temperature obtained smooth surfaces, reduced structural defects and improved the crystallinity of the metal oxide nanocluster. Owing to the high surface area of nanocluster, most of the atoms, dangling bonds, and local atomic defects exist at the surfaces texture of nanocluster rather than the inner cores. So, the surface of the nanocluster is unstable and has high chemical activity with a strong tendency to adsorb atoms from the surrounding environments. This active surface area is an essential factor in determining the detection limits or sensitivity. Aluri et al. have demonstrated highly selective and ultra-high sensitive sensors by functionalizing the TiO_2 nanoclusters on the GaN nanowire [23]. The deformation in the surface structure and the local defects acquire a vital role in the optical properties by creating energy bands in the band gap region which is an undesired scope in the field of optoelectronics application. Therefore, we expected that the annealing treatment can trigger the oxygen vacancies and improved the crystal quality of the nanostructure.

In this chapter, we will study the effect of annealing on different metal oxide nanoclusters for different applications. Besides reviewing the literature studied by the various groups, we will be focusing on the report from our research group.

2. Metal oxide nanocluster

2.1 ZnO nanocluster

ZnO is a striking n-type metal-oxide-semiconductor that has a wide bandgap of 3.37 eV, large exciton binding energy of 60 meV and excellent chemical stability, electrical, optical, piezoelectric and pyroelectric properties. ZnO nanostructures have gained tremendous interest due to their potential applications in various nanodevices such as nanogenerators, field emitters, UV detectors, gas sensors, and solar cells. Due to their advanced technological applications, high qualities of zinc oxide nanostructures are greatly demanded.

Xuehua et al. reported the co-doped ZnO nanocluster fabricated by using nanocluster-beam deposition and its effect of annealing at 400 and 700°C on the structural, optical and magnetic properties for diluted magnetic semiconductors (DMSs) [24]. The XRD pattern showed the enhanced crystalline structure by increasing the annealing temperature. A possible explanation is that as grown nanoclusters were highly porous after the random nucleation of atoms under nanocluster-beam deposition. However, after annealing in air, the crystallite sizes were increased from 6 to 20 nm with fewer defects due to the agglomeration of nearby nanoclusters from the obtained thermal kinetic energy. It was also found that the absorption spectra of the annealed film become steeper as compared to the prepared sample. A blue shift appeared at 368 nm (3.37 eV) in the as-grown film mainly due to the quantum confinement effects. However, after annealing at

400°C, a redshift at 365 nm (3.39 eV) observed due to the increased in the crystallite size induced by annealing. But, no change appeared as the annealing temperature increased to 700°C. The intensity has decreased after annealing in the range of 210–350 nm which indicate that the annealed sample attained more enhanced structure and crystallinity. A strong UV emission observed at 379 nm caused by the recombination of surface-bound acceptor-exciton complexes and the peak intensity of the UV emission became stronger after annealing due to the enhanced crystallinity. The film annealed at 400°C exhibited the highest saturation magnetization of 36 μemu. The improvement in saturation magnetization is due to the exchange of F-center mediation [25]. Thus, the results further suggested that the annealing at 400°C strongly affect ferromagnetic performance.

Recently, Aljawfi et al. [26] studied the impact of annealing on the structural and optical properties of ZnO nanoparticles through auto combustion route and annealed in air at different temperatures: 200, 400, 600, and 800°C. They showed that the film annealed at 200°C had a low-quality crystal. As the annealing temperature increased, the crystallite size increased to 37 nm from 13.8 nm. Due to annealing, the crystallites gained enough diffusion or activation energy and small crystallites migrate to relatively equilibrium sites in the crystal lattice with lower surface energy. The room temperature PL spectra with different annealing temperature were also studied. The sample annealed at low temperature (200°C) exhibited anomalous behaviors; where the UV-PL band evidently diminished that indicate the low crystal quality. The results also revealed the presence of clusters in the ZnO NPs associated with low annealing temperature, and the existence of clusters characterized by the diminished and blue shift in the UV band to lower wavelength. The increased in the intensity of UV-PL emission bands and the decreased in the intensity of the VIS-PL band in sequence with further annealing indicated the improvement in the crystallinity and reduction in the intrinsic local atomic defects. They also revealed that the annealing of ZnO clusters at or above 400°C optimized the crystallinity and enhanced the UV spectra related to optoelectronics application.

The heat treatment in air or pure oxygen may promote the reduction in oxygen vacancy and strong emission in ultraviolet band. Gao et al. further confirmed the above statement [27]. They have prepared the ZnO nanocluster porous films on a glass substrate by the successive ionic layer adsorption and reaction (SILAR) method and annealed at 400°C in Argon and air atmosphere for 2 h. A strong diffraction peak of (100) for the sample annealed in the air is obtained which implies that oxygen in the annealing atmosphere can enhance the crystalline properties of ZnO. The film annealed in air/Ar posses increased in the transmittance due to the reduced porosity. Moreover, the peak emission intensity at 380 nm decreased after annealing which indicated the high optical quality of the ZnO layer. The emission at 380 nm corresponds to the intrinsic transition of exciton from the conduction band to the valence band. Thus, they inferred that the annealing in air enhanced the crystalline properties of ZnO rather than annealing in Argon but have no evident effect on the photoluminescence of ZnO film.

We have carried out the fabrication of ZnO nanocluster and SiO_x/ZnO heterostructure nanocluster by using glancing angle deposition (GLAD) technique [28]. Glancing angle deposition (GLAD) technique is a physical vapor deposition technique for producing high-quality nanostructures as shown in **Figure 1**. The advantages of the GLAD technique are as follows: (i) the nanostructure is of high purity as they prepared in an ultra-high vacuum without any catalysts; (ii) the porosity of the nanostructure can be controlled by changing the incident angle; (iii) there is almost no restriction on materials since the growth process is thermal evaporation; (iv) the shape, and in-plane alignment of nanostructure can be easily modified; (v) self-alignment due to the shadowing effect directly on the device substrate

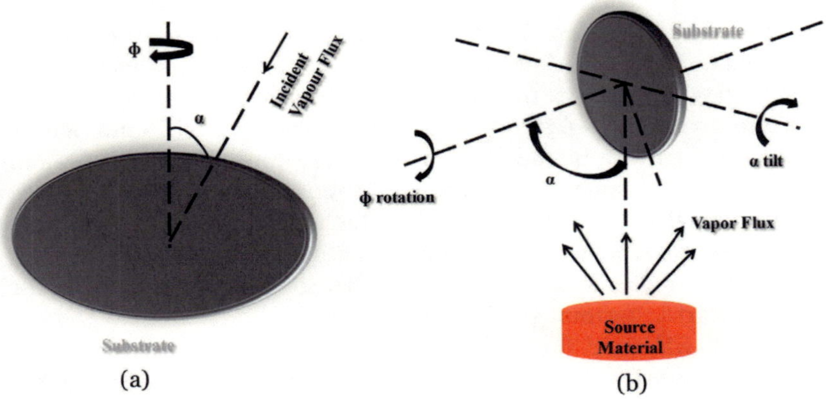

Figure 1.
(a) GLAD about incident angle α of the vapor flux and (b) substrate azimuthally (φ) rotation.

without the need for any additional experimental set up; and (vi) the technique could be used for commercial purposes.

GLAD can be used to engineer the nanostructure films with morphological features such as nanorods, nanowires, nanoclusters, helixes, zigzags, by adjusting the deposition angle, α and substrate rotation angle, φ. In this technique, three parameters determine the morphology of the nanostructure: the incident angle, the growth rate, and the substrate rotational speed. The high melting point materials have relatively low surface mobilities during the film growth processes. Therefore, during the growth process, a high melting point material flux is chosen to be incident onto the surface normal, and the substrate is rotating. GLAD produces nanostructure through the effect of shadowing during film growth with low surface mobilities force the adatoms to form porous nanostructures. Nanostructure films with their high surface to volume ratio are a potential application for optoelectronic devices since their unique morphology can enhance the device performance, namely: photosensitivity, wavelength selectivity, response and reset times, and reproducibility and stability.

The basic deposition setup is the same as that of oblique angle deposition is shown in **Figure 2**; the only difference is that the substrate manipulated by the incident angle Θ and the azimuthal rotation of the substrate concerning the substrate surface normal. By combining the oblique angle deposition and substrate positional control, a technique is introduced which is called glancing angle deposition (GLAD). In this deposition technique, the collimated evaporation beam has a large incident angle α concerning the substrate surface normal therefore incoming vapor can be considered as vector F. This vector F has two components, a vertical component $F_{\perp} = F\cos\alpha$ and α lateral component $F_{|} = F\sin\alpha$. The substrate will receive the vapor from both the components, but for the growth of the nanoclusters, the deposition should be preceded by $F_{\perp}$ only, i.e., cancel out the $F_{|}$ component. This forms the basis of the glancing angle deposition technique for the growth of nanoclusters. When the substrate rotates azimuthally, each part has an equal chance to receive the same amount of particle from the $F_{|}$ component. After a complete revolution, the average $\sum F_{|}$ is zero due to the cancelation of $F_{|}$ component at the opposite directions which means that there is no preferred orientation of the nanoclusters. By this technique purely nanoclusters will grow at the substrate surface.

The optical properties were carried out on both the samples by a UV–visible NIR spectrophotometer and photoluminescence spectroscopy (HITACHI, F-7000 FL spectrophotometer). Fourier transform infrared radiation methods (PERKIN ELMER, LAMBDA 900) were also carried out on both the samples. The average

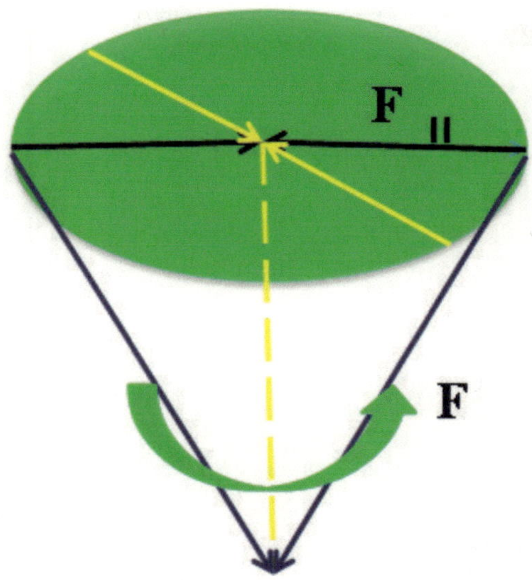

Figure 2.
The cancelation of the $F_{|}$ component due to the rotation. Only $F_{\perp}$ contributes effectively to the increasing height of the nanostructure.

crystallite size of the ZnO nanocluster and SiO_x/ZnO heterostructure nanocluster was found to be ~7.101 and ~10.14 nm respectively. The increase in crystallite size suggests the decrease in the existence of grain boundaries in the SiO_x/ZnO heterostructure nanocluster [29]. Moreover, the decrease in the average lattice strain observed from 0.01641 (ZnO nanocluster) to 0.01336 (SiO_x/ZnO heterostructure nanocluster).

The optical absorption spectrum showed in the wavelength range from 320 to 800 nm of the ZnO nanocluster and SiO_x/ZnO heterostructure nanoclusters in **Figure 3**.

ZnO is known to have a large number of defects like oxygen vacancies. When a photon generates an electron-hole pair, recombinations are more likely to take place in the oxygen vacancy. ZnO nanoclusters have a continuous absorption in the UV range as shown in **Figure 3**, demonstrating the presence of a noticeable amount of metallic zinc and a considerable amount of defects [30]. Some recent studies have shown that SiO_x has a conductive property. When electron-hole pairs initially generated in the ZnO nanocluster, the generated electrons accumulate in the conduction band, and they are attracted to the trap level of SiO_x [31]. Hence, due to the presence of SiO_x beneath the ZnO, an enhancement of the light absorption in the visible region beyond 400 nm as well as the near-infrared region can be observed.

The room temperature PL spectra of ZnO nanocluster and SiO_x/ZnO heterostructure nanocluster measured under 325 and 350 nm excitation (**Figure 4**). For comparison, PL spectrums showed for both the samples in the same figure. The excitation wavelength of 350 nm exhibits only a strong emission band in the blue region for both the samples attributed to near energy bandgap. The near band gap energy comes from the electron–hole recombination at a deep level (DL) emission in the band gap caused by intrinsic point defects and surface defects, e.g., oxygen vacancies, zinc interstitials. The blue emission peak at 450 nm (~ 2.75 eV) usually results from the radiative recombination of delocalized electrons close to the conduction band with deeply trapped holes in zinc vacancy. The intensity of blue emission under excitation wavelength 325 nm of ZnO nanocluster reduced slightly and exhibited non linear increase-decrease dependence, first increasing, the saturation at the bandgap energy as the optimal excitation energy and finally decreasing

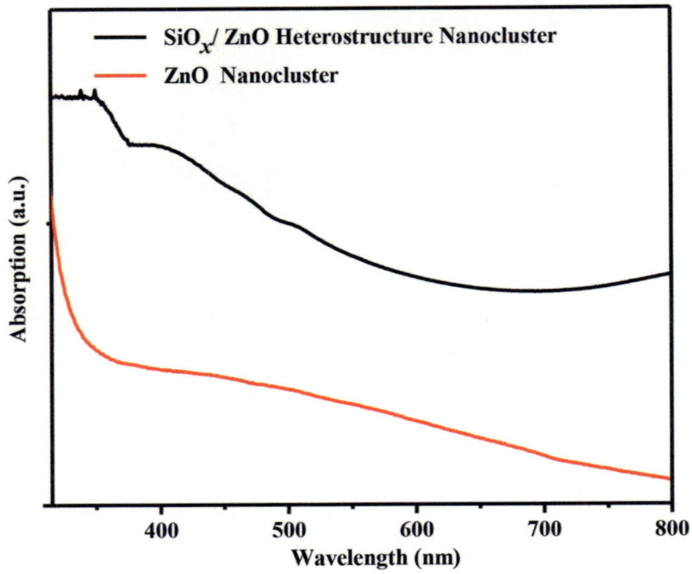

Figure 3.
The optical absorption spectrum of ZnO nanocluster and SiO_x/ZnO heterostructure nanocluster.

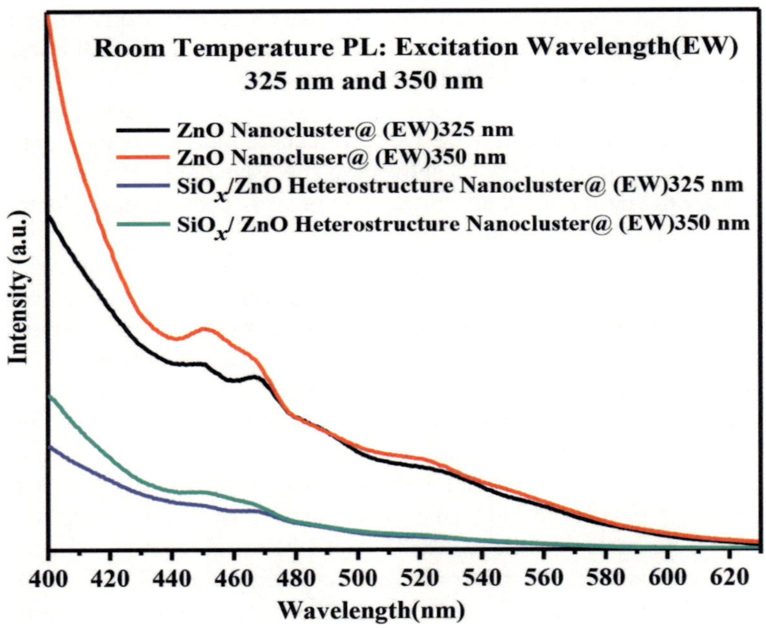

Figure 4.
Photoluminescence spectrum of ZnO nanocluster and SiO_x/ZnO heterostructure nanocluster.

but still effectively emitting as slowly dropping tail emission peaks at 467 nm is attributed to the transitions from Zn_i states to the valence band [32].

SiO_x/ZnO heterostructure nanocluster exhibited much lower emission intensity than ZnO nanocluster indicating that the recombination of the photogenerated charge carrier significantly inhibited in SiO_x/ZnO heterostructure nanocluster. The efficient charge separation may increase the lifetime of the charge carriers and enhance the efficiency of the interfacial transfer to adsorbed substrates and then account for the higher activity. Also, the peak shift from 450 to 457 nm in SiO_x/ZnO heterostructure nanocluster may be due to the radiative recombination of photo-excited electrons. These photoexcited electrons that accumulate in the conduction band of ZnO are attracted to the trap level of SiO_x due to the difference in work functions of ZnO and SiO_x [31].

FTIR spectra of ZnO nanocluster and SiO_x/ZnO heterostructure in the spectral range of 400 to 1500 cm^{-1} shown in **Figure 5** consisted of several bands at 420, 610, 738, 819 and 1109 cm^{-1}. The peak at 420, 610 and 738 cm^{-1} indicates the presence of ZnO which is in good agreement with the results obtained by XRD and attributed to the stretching vibration mode and also be related to the influence of shell induced strain and the appearance of ZnO bond between Zn and O atoms of the shell [33]. The spectra also show the bending vibration of Si—O—Si at 819 cm^{-1} indicating the bonding between the substrate and SiO_x. The peak located at 1109 cm^{-1} assigned to asymmetric Si—O stretching is due to interstitial oxygen impurities dissolved in the Si substrate. In SiO_x/ZnO heterostructure nanocluster, the entire peak transmittance % got quenched. Together these results identified the presence of SiO_x which exist near the ZnO surface.

The above work indicated that the SiO_x/ZnO heterostructure nanocluster could account for higher activity in optoelectronic devices. So, we further annealed the SiO_x/ZnO heterostructure nanocluster at 550°C for 1 h in open air [2]. The annealing treatment was carried out in the muffle furnace. The crystallinity improved after annealing owing to the decrease in the defect density. Also, the increased crystallite size attributed to the coalescence of grains. Moreover, the annealed sample becomes smoother as shown in **Figure 6** and offer a large surface area ascribed to the agglomeration of smaller nanocluster. The authors also confirmed that the oxygen content increased to 20.90% due to the adsorption of O_2 at the time of open-air annealing and hence, absorption is enhanced in the case of the annealed sample as shown in inset **Figure 6**.

The authors also reported the optical absorption spectrum in **Figure 7(a)** of the wavelength range from 320 to 900 nm of the bare ZnO nanocluster, as deposited and annealed SiO_x-ZnO heterostructure nanoclusters. An enhancement of averagely 1.5 times photon absorption is observed in the annealed sample with sharper band edge which indicated the formation of the improved crystal structure of the sample [30]. However, the decreased in the absorbance after annealing in the

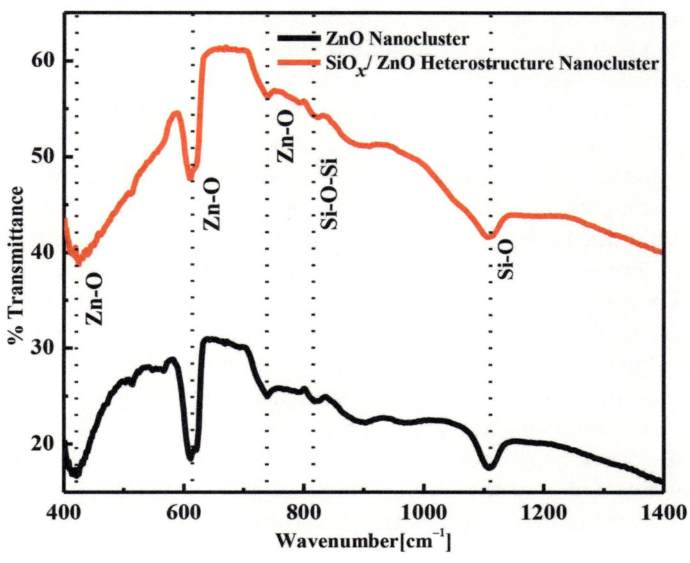

Figure 5.
FTIR spectra of ZnO nanocluster and SiO_x/ZnO heterostructure nanocluster.

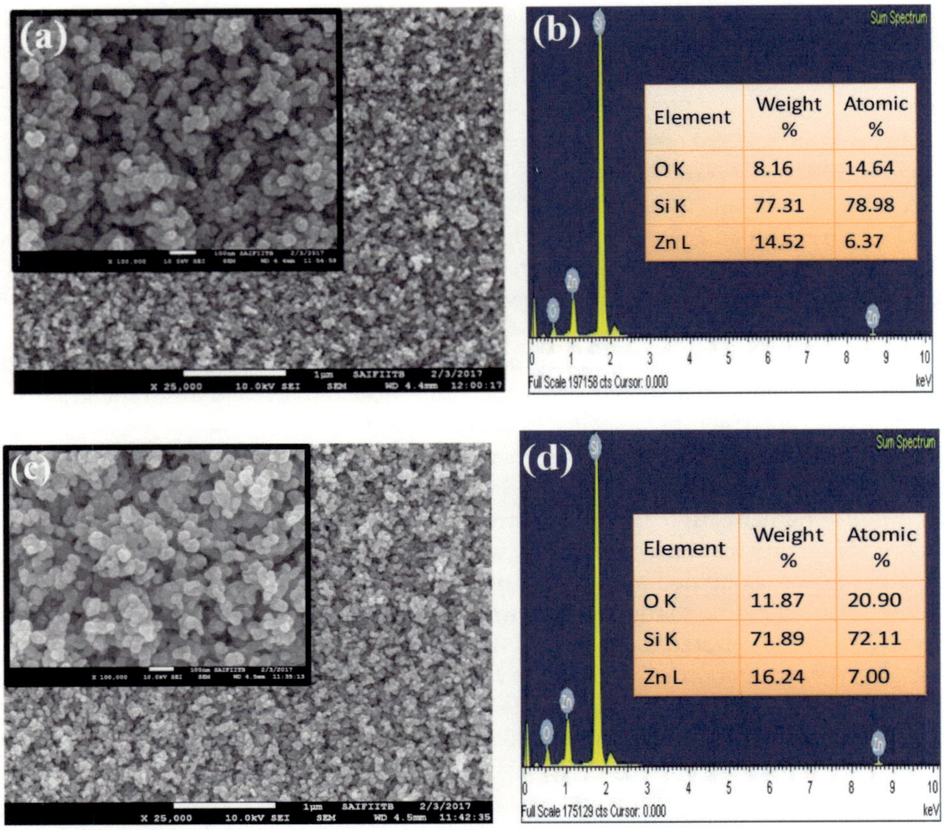

Figure 6.
FEG-SEM images of the SiO_x-ZnO heterostructure nanoclusters: (a) top view (as deposited); (b) EDAX analysis of as-deposited; (c) annealed and (d) top view (annealed).

wavelength range of ~374 to 478 nm is attributed to the change in the band gap. A continuous absorption is seen in the visible range of the as a deposited sample due to the noticeable amount of metallic zinc and a considerable amount of defects. The light absorption is enhanced in the visible region beyond 400 nm to NIR regions. Additionally, the voids were filled up with the O_2 after annealing which enhanced the probability of main band transitions and therefore the absorption in the UV region. Also, the subband transition of SiO_x shows the maximum absorption of at around 2.7 eV in the visible region. The interaction between Si-Si increased and formed the Si nanocluster in the SiO_x due to annealing. The absorption is in the near infrared region is due to the presence of Si. However, no light absorption beyond 850 nm was observed because Si absorbed light efficiently up to near infrared region only. That is why there is a steep change of absorption at 850 nm.

We have obtained the increase in the transmittance after annealing which attributed to the decreased in defects and an increase in the ZnO to Zn ratio [30]. The absorbed oxygen after annealing fills up the oxygen vacancies, hence reducing the density of this donor like defects. The main band gap of ZnO film is 3.37 eV [34]. However, the band gap of the as deposited SiO_x-ZnO heterostructure nanocluster is found to be 2.94 eV due to the downward shift of the conduction band. The less band gap associated with many-body interactions between the carriers in the conduction band and valence band called bandgap renormalization [35]. However, after annealing, the optical band gap increases to 3.44 eV which is influenced by the change in charge carriers and can be explained by Burstein Moss (BM) effect [35].

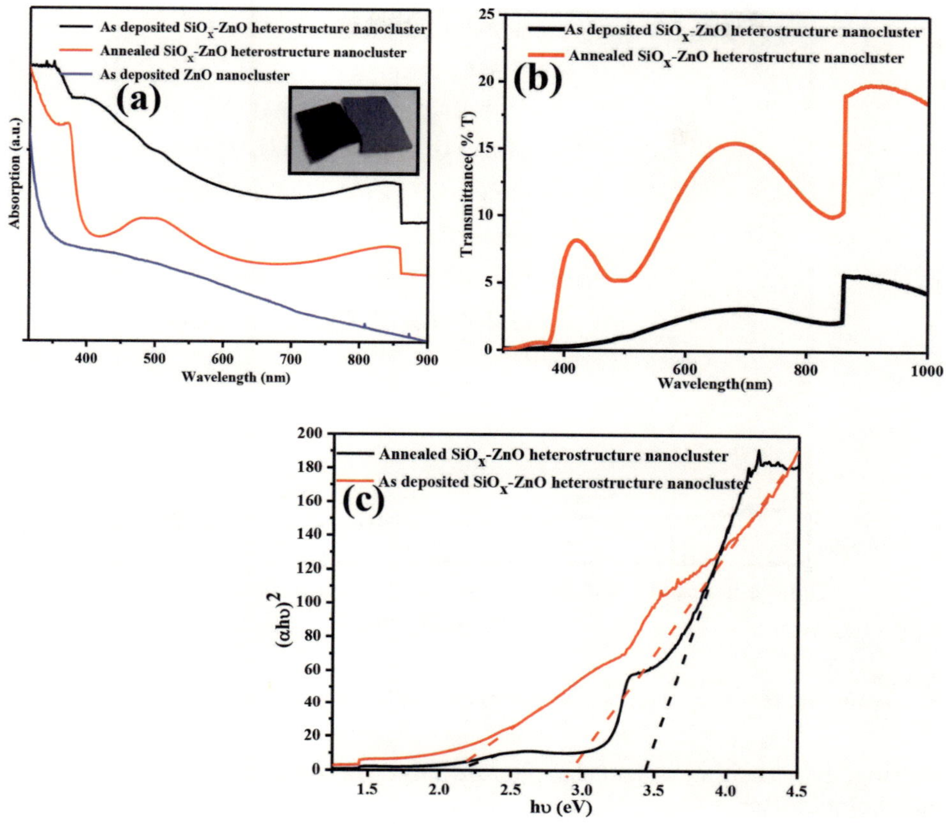

Figure 7.
(a) Optical absorption of as deposited ZnO nanocluster, as deposited and annealed SiO_x-ZnO heterostructure nanocluster and (b) transmittance (c) $(\alpha h\upsilon)^2$ versus $h\upsilon$ curve of as deposited and annealed SiO_x-ZnO heterostructure nanocluster.

The PL spectra of the as deposited and annealed SiO_x-ZnO heterostructure nanoclusters with different excitation wavelengths of 325, 350 and 375 nm under same 370 filter (**Figure 8**) also studied. The peak intensity significantly increased due to the formation of O_i and O_{Zn} after annealing at 550°C. Generally, the emission peak of ZnO nanostructure centered at around 415 nm. This emission peak arises due to the radiative defects in ZnO. However, in our case, the emission peak at around 420 nm is due to the radiative recombination of photoexcited electrons. These photoexcited electrons that accumulate in the CB of ZnO are attracted to the trap level of SiO_x due to the difference in work functions of ZnO and SiO_x. The intensity of blue emission exhibited a non linear increase-decrease dependence, first increasing, the saturation at the bandgap energy as the optimal excitation energy and finally decreasing but still effectively emitting as slowly dropping tail emission peaks at 450 nm and 468 nm which attributed to the transitions from Zn_i and extended Zn_i states to the valence band respectively. These extended states formed during the annealing process according to the defect ionization reaction [32].

2.2 TiO_2 nanocluster

Titanium dioxide (TiO_2) is a wide bandgap semiconductor with an indirect optical bandgap ranging from 3.2 eV for the anatase phase to 3.0 eV for the rutile phase. The unique physical, chemical, and optical properties have motivated more researchers in many applications such as sensors, photo-electrochemical water

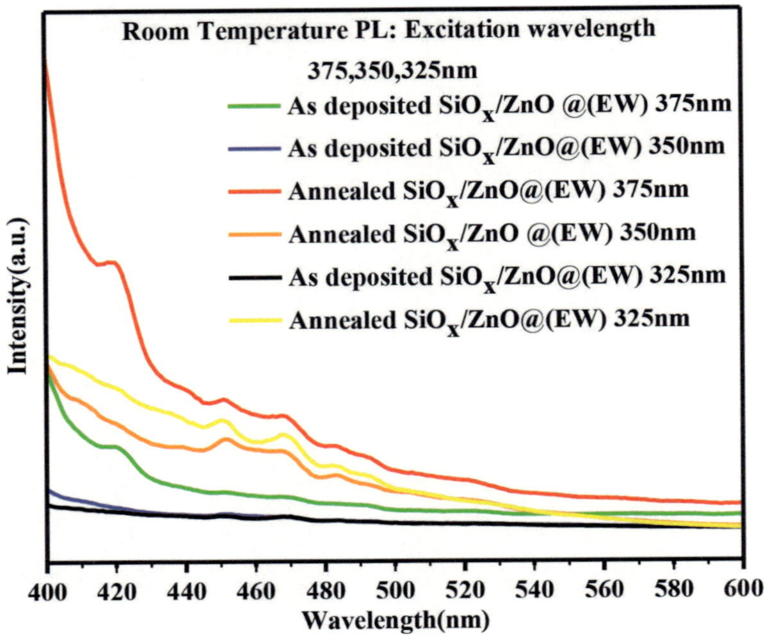

Figure 8.
Photoluminescence spectrum of as deposited and annealed SiO_x-ZnO heterostructure nanocluster.

splitting, dye-sensitized solar cells, photo-catalytic performance, memory devices, and photodetector.

The surface area of the nanostructure is one of the essential factors in determining the sensitivity of the chemical sensors. However, the surface/adsorbate interactions of the nanowires are often limited and non-specific. So, the functionalization of the NWs surface with metal/metal oxide nanoparticle or nanocluster focus at resolving the deficiencies of the NW-based devices. GS Aluri et al. demonstrated GaN nanowires decorated with TiO_2 nanoclusters for selectively sensing benzene and related aromatic environmental pollutants such as toluene, ethylbenzene and xylene [23]. The TiO_2 nanoclusters act as catalysts design to increase the sensitivity, lower the detection time since nanoclusters/nanoparticles can increase the adsorption of chemical species by introducing additional adsorption sites. Herein, TiO_2 nanoclusters fabricated by using RF magnetron sputtering followed by an annealing at 650–700°C in a rapid thermal processing system of ultrahigh purity Ar. The TiO_2 were amorphous in as-deposited film. Upon annealing of TiO_2 nanoclusters at 700°C for 3 s, the phase changed to anatase in the TiO_2 clusters. The I-V curve was nonlinear and asymmetric in the as a deposited sample but the current increased with the deposition of TiO_2 nanoclusters and significant change occurred when the device annealed at 700°C. They have also shown that the photoconductivity increased almost two orders of magnitude for the device decorated with TiO_2 nanoclusters. The enhancement of photoconductance after decorating nanoclusters to nanowires is attributed to the separation of photogenerated charge carriers by a surface depletion field, thereby increasing the lifetime of the photogenerated carriers. Thus, these results demonstrated the potential of hybrid chemical sensors using TiO_2 nanoclusters.

Wei et al. have prepared the *R*-TiO_2 clusters by post-deposition oxidation of Ti clusters with rapid thermal annealing in an oxygen atmosphere [36]. The *R*-TiO_2 clusters showed the highest crystalline structure as compared to Ti, TiO, and A-TiO_2. In a different work, Ag-TiO_2 nanocomposite was prepared by DC reactive magnetron sputtering followed by vacuum annealing treatment from 200 to 800°C.

The friction coefficient increased from 0.21 to 0.7 after the sample annealed at 800°C [37]. Recently, Song et al. fabricated the TiO_2 nanocluster-SiO_x (x˂2) using atomic layer deposition (ALD) for the properties of resistive switching. They performed thermal annealing in N_2 ambient at 900°C for 3 h using a thermal furnace. They have reported the current on/off ratio of the device is about 10^5 (reading at +1.0 V), which is much higher than that of many Resistance change random access memories (RRAM) devices of TiO_2 [38].

TiO_2 nanocluster determined the photocatalytic activity by its crystalline phase (anatase and rutile), crystallite size, specific surface area, pore structure, and crystallinity. Khan et al. reported that the crystallization of TiO_2 film to anatase crystal phase starts occurring at the annealing temperature of 300–600°C in the air which improves the crystal structure of the TiO_2 anatase [39]. Bin et al. investigated the effect of GLAD on TiO_2 nanostructure with different annealing temperature for the morphology, and crystallization and PL. The average size increased with the increase in annealing temperature, and defects were reduced [40].

Min Liu et al. reported the photochemically inactive Ti3+ self-doped TiO_2 prepared by a facile one-step thermal oxidation into an efficient visible light-sensitive photocatalyst by the grafting of Cu (II) oxide amorphous nanoclusters [41]. The grafting of Cu (II) oxide onto TiO_2@Ti3+ performed by the impregnation method shown that the grafting of Cu (II) nanoclusters to TiO_2 increased the absorption intensities in the 420–550 and 700–800 nm wavelength regions assigned to the interfacial charge transfer (IFCT) VB electrons to surface-grafted Cu (II) nanoclusters. On the contrary, the grafting of Cu (II)-TiO_2@Ti3+ the absorption intensities decreased slightly in the range of 420–550 nm wavelength region, while the increased in the absorption intensities can also be observed in the range of 700–800 nm wavelength region attributed to the d-d transition of Cu(II). They have shown that the TiO_2@ Ti3+ had a negligible activity either under visible light or UV light irradiation without the grafting of nanocluster. After grafting, they have switched to an efficient photocatalyst. Further, they have studied the photocatalytic activities of Cu (II)-TiO_2@Ti3+ under visible light irradiation at 400–530 nm (1 mW/cm^2) with different annealing temperature. However, the ratio of TiO_2 to Ti_2O_3 greatly influenced the morphology and the degree of oxidation states rather than annealing.

2.3 Tungsten oxide nanocluster

In the family of metal transition oxides, tungsten trioxide (WO_3) have attracted due to their distinctive properties such as chromogenic properties, gas sensitivity, and photosensitivity. Tungsten oxide nanoclusters were fabricated using supersonic cluster beam deposition (SCBD) by Zhao et al. [21] for examining the H_2 sensing properties. Further, the sample was post-annealed at 250°C for 12 h. They showed a large sensor signal of 1200 at 3000 ppm of H_2 in the air due to the small cluster size. The response of H_2 was over a broad range from 4 to 20,000 ppm. Moreover, the detection limit is as low as 0.042 ppm. Lastly, the sensor exhibit fast response rate.

2.4 Iron oxide nanocluster

The unique properties such as superparamagnetism arise from the fluctuation in the direction of the magnetic moment in a single domain nanoparticle due to thermal agitation. Marin et al. reported a novel core-shell superparamagnetic iron oxide nanocluster under annealing treatment [42]. The annealing of iron oxide nanocluster at 300°C for 3 h in air greatly influences the magnetic properties. Due to the thermal decomposition of the surfactants during the annealing treatment and in the absence of oxygen inside the nanoclusters led to the formation of an amorphous carbon" coat" around the maghemite nanoparticle. After the annealing treatment, the value of saturation magnetization is increased by about ~13.6%. This increase in the magnetization of the annealed sample can be attributed to the degradation and decomposition of the surfactants and their release of gas products during the annealing treatment from the sample, i.e. to the higher content of the magnetic material in the annealed sample. They showed the decrease in the interactions in the annealed sample indicated that the maghemite nanoparticles are better isolated and shield each other with the amorphous carbon.

Apart from the nanoclusters, the effect of annealing on metal oxide nanostructures such as indium oxide and tin oxide will be discussed in the next section.

2.4.1 Indium oxide

Indium oxide (In_2O_3), known as an n-type, is a wide-bandgap transparent semiconductor (with a direct bandgap of ~3.6 eV and an indirect bandgap of ~2.5 eV) is of great interest for diverse technological applications in nanoelectronics and optoelectronics. Zero-dimensional In_2O_3 nanoparticles/nanoclusters, with a variety of tunable, are beneficial as building blocks for indium oxide-based or hybrid transistor. Their remarkably surface-to-volume ratio and good stability have made them promising materials in gas sensors/biosensors, photocatalysis, photoelectrochemical cells, and ultraviolet photodetectors. Despite the advantages, In_2O_3 nanostructure usually encounters low conductivity, and this could decrease the stability and efficiency of the device. Due to the weak interconnection between the nanoparticles/nanoclusters, the carrier transportation between the NPs lost at the interface due to the charge delocalization. One way to solve this problem is to improve the structure by annealing and plasma treatment.

Kong et al. reported the zero-dimensional In_2O_3 nanoparticle sample showed an improvement in the crystallinity and a more compact structure after the thermal treatment [43]. It also showed an improvement in the optical and electrical properties of the In_2O_3 nanoparticle. Flores et al. studied the effect of annealing on the electrical, optical and structural properties of the undoped In_2O_3 nanostructure [44]. The film first annealed at 300°C in the air and the second annealing at 500°C results with a better passivation effect of the traps, whether located at grain boundaries or in the bulk of the indium oxide grains. So, the highest carrier concentration, mobilities, and band gap value obtained for the samples treated under 300 and 500°C. Sudha et al. [45] also investigated the crystallinity of In_2O_3 nanostructures along with the grain size augmented with annealing temperature. They observed that average grain size increased with increasing annealing temperature and explained by considering the thermal annealing-induced coalescence of small grains by grain boundary diffusion which caused significant grain growth. Microstrain and dislocation density present in the In_2O_3 nanostructure are found to decrease due to thermal annealing implies that the stress is compressive which is likely to be governed by the dominant crystallization process. Senthilkumar et al. presented the influence of annealing temperature on the structural and optical properties of the In_2O_3 films deposited by electron beam evaporation technique in the presence of oxygen [46]. They have also proved that the crystallinity of In_2O_3 films improved with annealing temperature. The optical band gap is found to vary from 3.65 to 3.86 eV with increasing annealing temperature. The shift in the band gap to higher energies attributed to the increase in carrier density which arises from the filling of states near the bottom of the lowest state in the conduction band. The results also show that by increasing the annealing temperature, the electrical resistivity of In_2O_3 thin films decreases which is related to the increase of the mobility and carrier density. Thus, the annealing temperature plays a vital role in the conductivity of the In_2O_3 nanostructure as well.

2.4.2 Tin oxide

In recent years SnO_2 has attracted a lot of interest because of its outstanding electrical, optical and electrochemical properties and these properties enabled the application of SnO_2 in solar cells, catalytic support materials, transparent electrodes, and solid-state chemical sensors. Among the various phases of tin oxides, SnO is the most metastable phase, which is mostly formed at high annealing temperatures (450 and 750°C). The size, morphology, and phase of the tin oxide nanoparticles can be effectively modified by the process of annealing. The effect of pH and annealing temperature on the properties of tin oxide nanoparticles prepared by the sol–gel method has reported [47]. The samples were annealed at 200, 400, 600, and 800°C. The average crystallite size increased with the increase in annealing temperature. The surface morphology of the tetragonal SnO_2 nanoparticles studied by scanning electron microscope revealed the formation of spherically shaped agglomerations.

On the other hand, carbon nanomaterials composed of sp^2 bonded graphitic carbon are found in different dimensions including zero-dimensional fullerenes, one-dimensional carbon nanotubes (CNTs), and two-dimensional graphene. Owing to the unique structure and physical properties, carbon-based nanomaterials have gained lots of interests in electronic, optoelectronic, photovoltaic, and sensing applications. Among the different carbon nanomaterials, CNTs provides unique opportunities for novel optoelectronic properties since semiconducting CNTs are direct bandgap materials that possess free electron-hole pair excitations as well as strongly bound electron-hole pair states called excitons. One dimensional nature of CNTs produces van Hove singularities in the density of states that result in strong optical absorption and emission with energies. The use of graphene for optoelectronic devices is limited due to its zero band gap at the Fermi level. However, by tuning the band gap, graphene can be an ideal candidate for optoelectronic devices [48]. CNTs are also chemically inactive due to their strong sp^2 bonding in a near perfect hexagonal network and this prevents the formation of chemical bonds with most molecules. Its reactivity can be improved towards different gases by decorating the CNTs wall by metal or metal oxide nanocluster due to its physico-chemical properties (e.g. high catalytic activity, adsorption capacity, efficient charge transfer, etc.). Nanoclusters play a major role in the gas detection pathway, in which sensitivity and selectivity can be tuned based on the reactivity of the nanoclusters surfaces and on the nature of the charge transfer between carbon nanotubes and nanoclusters induced by gas adsorption [12]. In addition, the incorporation of metal nanoclusters (Ag, Au) inside carbon host matrix can add functionality to carbon films. However, the high internal stress of carbon can affect the structural evolution of metal incorporated carbon film. Jose et al. [49] studied the influence of annealing in argon at 300°C on the conductivity, phase stability and electronic structure of hydrogen-free amorphous carbon (a-C) films containing copper (aC:Cu) and gold (a-C:Au) nanoclusters. They showed that the a-C host matrix increased its graphitic character and stress was relieved upon annealing due to increase in the high energy side of the σ^* region. Also, the structural transformation after annealing the carbon incorporated nanoclusters changes their electronic conductivity and apparent optical bandgap.

3. Conclusions

The chapter reviewed the effect of annealing on ZnO, SiO_x, and TiO_2 nanocluster for different applications. ZnO, SiO_x, and TiO_2 have been fabricated using different physical and chemical method. Our work demonstrated the influence of annealing on optoelectronics properties on GLAD synthesized SiO_x-ZnO heterostructure nanoclusters. The post-annealing treatment in oxygen ambient improved the structural, optical and electrical properties. Nanocluster based devices outperform their bulk counterparts. In addition, some different metal oxide nanostructure and its effect of annealing also discussed. Therefore, we may conclude that future nanodevices based on nanocluster should be paid attention to the industrial and commercial applications.

Acknowledgements

The authors would like to thank especially the initiator of this book for making them a part of this vital endeavor.

Conflict of interest

The author declares that there is no conflict of interests regarding the publication of this book chapter.

Appendices and nomenclature

V_{Zn}	zinc vacancy
O_i	oxygen interstitial
O_{Zn}	antisite oxygen
Zn_i	interstitial zinc ion

Author details

Naorem Khelchand Singh* and Rajshree Rajkumari
Department of Electronics and Communication Engineering, National Institute of Technology, Nagaland, Dimapur, India

*Address all correspondence to: khelchand.singh@gmail.com

References

[1] Liu H, Yang W, Wang M, Xiao L, Liu S. Fabrication of lotus-like Au@ TiO_2 nanocomposites with enhanced gas sensing properties. Sensors and Actuators B. 2016:236-490. DOI: 10.1016/j.snb.2016.06.039

[2] Rajkumari R, Singh NK. Influence of annealing on the optoelectronic properties of the GLAD synthesized SiO_x–ZnO heterostructure nanoclusters. Applied Physics A: Materials Science & Processing. 2018:124-264. DOI: 10.1007/s00339-018-1687-1

[3] Frederick RT, Novotny Z, Netzer FP, Herman GS, Dohnalek Z. Growth and stability of titanium dioxide nanoclusters on graphene/Ru (0001). The Journal of Physical Chemistry. B. 2018:122-640. DOI: 10.1021/acs.jpcb.7b05518

[4] Lu Y, Xu YJ, Zhang GB, Ling D, Wang MQ, Zhou Y, et al. Iron oxide nanoclusters for T 1 magnetic resonance imaging of non-human primates. Nature Biomedical Engineering. 2017;**1**(8):637-643. DOI: 10.1038/s41551-017-0116-7

[5] Dinh LN, Schildbach MA, Balloch M, McLean IIW. Pulsed laser deposition of ZnO nanocluster films by Cu-vapor laser. Journal of Applied Physics. 1999:86-1149. DOI: 10.1063/1.370857

[6] Slaughter G, Sunday J. Fabrication of enzymatic glucose hydrogel biosensor based on hydrothermally grown ZnO nanoclusters. IEEE Sensors Journal. 2014:14-1573. DOI: 10.1109/JSEN.2014.2298359

[7] Li X, Zhao Z. Co-doped ZnO thin films fabricated by a nanocluster-beam deposition system and the influence of flow rate of helium gas on their properties. Journal of Superconductivity and Novel Magnetism. 2016;**29**:357. DOI: 10.1007/s10948-015-3230-3

[8] Wu S, Yuan N, Xu H, Wang X, Schelly ZA. Synthesis and bandgap oscillation of uncapped, ZnO clusters by electroporation of vesicles. Nanotechnology. 2006:17-4713. DOI: 10.1088/0957-4484/17/18/031

[9] Liu M, Inde R, Nishikawa M, Qiu X, Atarashi D, Sakai E, et al. Enhanced photo activity with nanocluster-grafted titanium dioxide photocatalysts. ACS Nano. 2014:8-7229. DOI: 10.1021//nn5 10.1021/nn502247x

[10] Yeow SC, Ong WL, Wong ASW, Ho GW. Template-free synthesis and gas sensing properties of well-controlled porous tin oxide nanospheres. Sensors and Actuators B: Chemical. 2009:143-295. DOI: 10.1016/j.snb.2009.08.050

[11] Lassesson A, Schulze M, Lith JV, Brown SA. Tin oxide nanocluster hydrogen and ammonia sensors. Nanotechnology. 2008:19-015502. DOI: 10.1088/0957-4484/19/01/015502

[12] Leghrib R, Pavelko RG, Felten A, Vasiliev A, Cané C, Gràcia I, et al. Gas sensors based on multiwall carbon nanotubes decorated with tin oxide nanoclusters. Sensors and Actuators B: Chemical. 2010:145-411. DOI: 10.1016/j.snb.2009.12.044

[13] Salavati-Niasari M, Mir N, Davar F. A novel precursor for the synthesis of metallic copper nanocrystals by thermal decomposition approach. Applied Surface Science. 2010:256-4003. DOI: 10.1016/j.apsusc.2010.01.067

[14] Amoli V, Farooqui S, Rai A, Santra C, Rahman S, Sinha A, et al. Indium oxide nanocluster doped TiO_2 catalyst for activation of molecular O_2. RSC Advances. 2015;**5**. DOI: 10.1039/C5RA13104A

[15] Amoli V, Malayil MGS, Banerjee B, Anand M, Maurya A, Farooqui S, et al.

Faceted titania nanocrystals doped with indium oxide nanoclusters as a superior candidate for sacrificial hydrogen evolution without any noble-metal Co catalyst under solar irradiation. ACS Applied Materials & Interfaces. 2014;7. DOI: 10.1021/am507293b

[16] Walsh A, Woodley S. Evolutionary structure prediction and electronic properties of indium oxide nanoclusters. Physical Chemistry Chemical Physics. 2010:12-8446. DOI: 10.1039/c0cp00056f

[17] Ayesh A I, Abu Hani A, Mahmoud S, Haik Y. Selective H_2S sensor based on CuO nanoparticles embedded in organic membranes. Sensors and Actuators B: Chemical. 2016:231. DOI: 10.1016/j.snb.2016.03.078

[18] Yin G, Nishikawa M, Nosaka Y, Nagarajan S, Atarashi D, Sakai E, et al. Photocatalytic carbon dioxide reduction by copper oxide nanocluster-grafted niobate nanosheets. ACS Nano. 2015: 9-2111. DOI: 10.1021/nn507429e

[19] Lomnicki S, Wu H, Osborne NS, Pruett J, M, McCarley RL, Poliakoff E, et al. Size-selective synthesis of immobilized copper oxide nanoclusters on silica. Materials Science and Engineering: B. 2010:175-136. DOI: 10.1016/j.mseb.2010.07.016

[20] Tiwari A, Nagaiah TC, Bordoloi A. Electrocatalytic activity of tungsten oxide nanoclusters grafted on mesoporous nitrogen-rich carbon material in the dioxygen reduction reaction. ChemPlusChem. 2015;**80**. DOI: 10.1002/cplu.201500253

[21] Zhao M, Xing Huang J, Wo Ong C. Room-temperature resistive H_2 sensing response of Pd/WO_3 nanocluster-based highly porous film. Nanotechnology. 2012:23-315503. DOI: 10.1088/0957-4484/23/31/315503

[22] Wegner K, Piseri P, Tafreshi H, Milani P. Cluster beam deposition: A tool for nanoscale science and technology. Journal of Physics D: Applied Physics. 2006;**39**:439. DOI: 10.1088/0022-3727/39/22/R02

[23] Aluri GS, Motayed A, Davydov AV, Oleshko VP, Bertness KA, Sanford KA, et al. Highly selective GaN-nanowire/TiO_2-nanocluster hybrid sensors for detection of benzene and related environment pollutants. Nanotechnology. 2011:22-295503. DOI: 10.1088/0957-4484/22/29/295503

[24] Li X, Zhao Z. Effect of annealing on Co-doped ZnO thin films prepared by nanocluster-beam deposition. Journal of Superconductivity and Novel Magnetism. 2016;**29**:1897. DOI: 10.1007/s10948-016-3395-4

[25] Zhao ZW, Tay BK, Chen JS, Hu J, Sun X, Tan ST. Optical properties of nanocluster-assembled ZnO thin films by nanocluster-beam deposition. Applied Physics Letters. 2005:87. DOI: 10.1063/1.2149170

[26] Aljawfi R, Alam M, Rahman F, Ahmad S, Shahee A, Kumar S. Impact of annealing on the structural and optical properties of ZnO nanoparticles and tracing the formation of clusters via DFT calculation. Arabian Journal of Chemistry. 2018. DOI: 10.1016/j.arabjc.2018.04.006

[27] Gao X. Synthesis and optical properties of ZnO nanocluster porous films deposited by a modified SILAR method. Applied Surface Science. 2004;**229**:275-281. DOI: 10.1016/S0169-4332(04)00104-7

[28] Robbie K, Brett M. Method of depositing shadow sculpted thin films. US Patent 5. 1999:866-204

[29] Akgul G, Akgul FA, Unalan HE, Turan R. Photovoltaic performance of gallium-doped ZnO thin film/Si nanowires heterojunction diodes. Philosophical Magazine. 2016:96-1093. DOI: 10.1080/14786435.2016.1154207

[30] Agarwala DC, Chauhan RS, Kumar A, Kabiraj D, Singh F, Khan SA, et al. Synthesis and characterization of ZnO thin film grown by electron beam evaporation. Journal of Applied Physics. 2006:99-123105. DOI: 10.1063/1.2204333

[31] Dhar JC, Mondal A, Singh NK, Chakrabartty S, Bhattacharyya A, Chattopadhyay KK. Effect of annealing on SiO_x-TiO_2 axial heterostructure nanowires and improved photodetection. Journal of Applied Physics. 2013:114-244310. DOI: 10.1063/1.4858420

[32] Zeng H, Duan G, Li Y, Yang S, Xu X, Cai W. Blue luminescence of ZnO nanoparticles based on non-equilibrium processes: Defect origins and emission controls. Advanced Functional Materials. 2010:20-561. DOI: 10.1002/adfm.200901884

[33] Raevskaya AE, Panasiuk YV, Stroyuk OL, Kuchmiy SY, Dzhagan VM, Milekhin AG, et al. Spectral and luminescent properties of ZnO–SiO_2 core-shell nanoparticles with size-selected ZnO cores. RSC Advances. 2014;**4**:63393. DOI: 10.1039/c4ra07959k

[34] Rodnyi A, Khodyuk IP. Optical and luminescence properties of zinc oxide. Optics and Spectroscopy. 2012;**111**. DOI: 10.1134/S0030400X11120216

[35] Jeon JW, Jeon DW, Sahoo T, Kim M, Baek JH, Hoffman JL, et al. Effect of annealing temperature on optical band-gap of amorphous indium zinc oxide film. Journal of Alloys and Compounds. 2011:509-10062. DOI: 10.1016/j.jallcom.2011.08.033

[36] Wei X, Skomski R, Balasubramanian B, Sun ZJ, Sellmyer D. Magnetism of TiO and TiO_2 clusters. Journal of Applied Physics. 2009: 105-07C517. DOI: 10.1063/1.3074509

[37] Munteanu D, Torrell M, Cunha L, Alves E, Barradas N, Cavaleiro A, et al. The influence of annealing treatments on the properties of Ag:TiO_2 nanocomposite films prepared by magnetron sputtering. Applied Surface Science. 2012:258-4028. DOI: 10.1016/j.apsusc.2011.12.095

[38] Song S, Kim K, Ho Jung K, Sok J, Park K. Properties of resistive switching in TiO_2 nanocluster-SiO_x (X<2) matrix structure. Journal of Semiconductor Technology and Science. 2018:18-108. DOI: 10.5573/JSTS.2018.18.1.108

[39] Khan Abdul F, Mehmood M, Durrani S, L, Ali M, Abd Rahim N. Structural and optoelectronic properties of nanostructured TiO_2 thin films with annealing. Materials Science in Semiconductor Processing. 2014:29. DOI: 10.1016/j. mssp.2014.02.009i

[40] Wang B, Qi H, Chai Y, Li M, Guo M, Pan M, et al. Alteration of titanium dioxide material properties by glancing angle deposition plus annealing treatment. Superlattices and Microstructures. 2016:90-87. DOI: 10.1016/j.spmi.2015.12.007

[41] Liu M, Qiu X, Miyauchi M, Hashimoto K. Cu(II) oxide amorphous nanoclusters grafted Ti3+ self-doped TiO2: An efficient visible light photocatalyst. Chemistry of Materials. 2011:23-10. DOI: 1021/cm203025b

[42] Tadic M, Kralj S, Jagodic M, Hanzel D, Makovec D. Magnetic properties of novel superparamagnetic iron oxide nanoclusters and their peculiarity under annealing treatment. Applied Surface Science. 2014:322-255. DOI: 10.1016/j.apsusc. 2014.09.181

[43] Chong Su K, Azizan SNA, Wah C, Nguyen H, Siong Chiu W, Aspanut Z, et al. Structure deformation of indium oxide from nanoparticles into nanostructured polycrystalline films by in situ thermal radiation treatment. Nanoscale Research Letters. 2013:8. DOI: 10.1186/1556-276X-8-428

[44] Flores MM, Castanedoperez R, Torresdelgado G, Marquezmarin J, Zelayaangel O. Influence of the annealing temperature on the properties of undoped indium oxide thin films obtained by the sol-gel method. Thin Solid Films, DOI. 2008:517-681. DOI: 10.1016/j.tsf.2008.07.036

[45] Sudha A, Sharma SL, Maity TK. Effects of annealing temperature on structural and electrical properties of indium oxide thin films prepared by thermal evaporation. Materials Letters. 2015:157. DOI: 10.1016/j.matlet.2015.05.050

[46] Velusamy S, Vickraman P. Annealing temperature dependent on structural, optical and electrical properties of indium oxide thin films deposited by electron beam evaporation method. Current Applied Physics. 2010;**10**:880-885. DOI: 10.1016/j.cap.2009.10.014

[47] Priya M, Arunachalam G, Ramamurthi K, Veluswamy P, Hiroya I. Effect of pH and annealing temperature on the properties of tin oxide nanoparticles prepared by sol–gel method. Journal of Materials Science: Materials in Electronics. 2017:29-658. DOI: 10.1007/s10854-017-7959-2

[48] Shahrokhi M. Tuning the band gap and optical spectra of monolayer penta-graphene under in-plane biaxial strains. Optik. 2017:205-214. DOI: 10.1016/j.ijleo.2017.02.033

[49] Endrino JL, Horwat D, Anders A, Andersson J, Gago R. Impact of annealing on the conductivity of amorphous carbon films incorporating copper and gold nanoparticles deposited by pulsed dual cathodic arc. Plasma Processes and Polymers. 2009:S438-S443. DOI: 10.1002/ppap.200931003